The Complete Book of Comprehensives

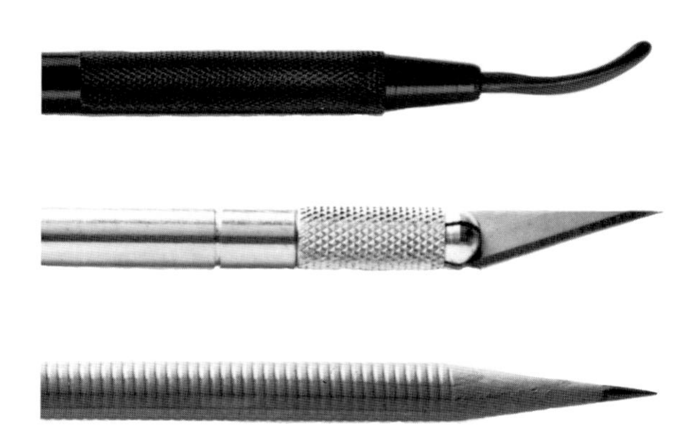

Robert Cullinane

VNR VAN NOSTRAND REINHOLD
New York

Library of Congress Catalog Card Number 89-5498
ISBN 0-442-21742-0

Printed in the United States of America

Designed by Keano Design Studio

Van Nostrand Reinhold
115 Fifth Avenue
New York, New York 10003

Van Nostrand Reinhold International Company Limited
11 New Fetter Lane
London EC4P 4EE, England

Van Nostrand Reinhold
480 La Trobe Street
Melbourne, Victoria 3000, Australia

Nelson Canada
1120 Birchmount Road
Scarborough, Ontario, M1K 5G4, Canada

16 15 14 13 12 11 10 9 8 7 6 5 4 3 2 1

Library of Congress Cataloging-in-Publication Data

Cullinane, Robert, 1958–
The complete book of comprehensives/Robert Cullinane.
p. cm.
Bibliography: p.
Includes index.
ISBN 0-442-21742-0
1. Graphic arts—Handbooks, manuals, etc. 2. Graphic arts—Technique. 3. Commercial art—Handbooks, manuals, etc. 4. Commercial art—Technique. I. Title.
NC997.C76 1990 89-5498
741.6—dc19 CIP

See p. 185 for trademark information

Contents

Introduction vii

CHAPTER 1 THE DESIGNING PROCESS 1

CHAPTER 2 BASIC STEPS IN THE GRAPHIC DESIGN PROCESS 3

Steps in Solving Graphic Design Problems 3
Research and Analysis 4
Sketching Thumbnails 6
Preparing Rough Layouts 7
Preparing Finished Layouts 9
Preparing Comprehensives 10
Preparing Advanced Comprehensives 13

CHAPTER 3 EQUIPMENT, TOOLS, AND MATERIALS 19

Equipment 19
Chairs and Stools 19
Drawing Boards 20
Drawing Tables 20
Taborets 20
Rotating Trays 21
Lamps 21
Light Boxes 21
Loupe 21
Hair Dryers 22
Lucy 22
Exposure Units 23
Copying Machines 23

Photostat Machines 23
Opaque Projectors 24
Condensing and Expanding Lens 24
Tools 25
Rules and Scales 25
Drawing Guides 27
Drawing Instruments 29
Cutting Tools 31
Brushes and Color Applicators 32
Pencil- and Lead-Sharpening Tools 35
Tape Dispensers 36
Burnishers 36
Color Specifiers 36
Note Pads 37
Materials and Media 38
Pencils 38
Erasers 39
Pens 40
Inks 42
Color Media 43
Papers and Pads 45
Art Boards 47
Acetates and Film Overlays 47
Tapes 49
Adhesives 50
Spray Coatings and Fixatives 55
Wipes and Cleaning Aids 56
Presentation Materials 56

CHAPTER 4 BASIC TECHNIQUES IN THE GRAPHIC DESIGN PROCESS 59

Making a Chisel Point 59
Squaring Up 60
Making Templates 61
Brush Ruling 61
Drawing with Liquid Media 61
Indicating Text Type 61
Text Type Indication Guidelines 62
Ruled-Line Method of Indicating Text Type 63
Loop Method of Indicating Text Type 65
Indicating Display Type 67
Pencil Lettering 67
Ink Lettering 69
Color Lettering 69
Color Marker Lettering 70
Cutout Lettering 70
Rendering Type on Acetate Overlays 71
Transferring Techniques 71
Graphite Transfer Method 72
Light Box Tracing 72
Enlarging Thumbnail Sketches 72
Visual Estimation Techniques 73
Mechanical Enlarging 73
Photographic Enlarging 73

Scaling Artwork 73
Diagonal Line Scaling 73
Using a Circular Proportional Scale 74
Joining Materials 74
Butting 74
Insetting 76
Preparing Interim Mechanicals 77
Preparing Interim Mechanicals with Several Colors 77
Altering Type 78
Altering Line Spacing 78
Altering Letter Spacing 79
Curving Type 80
Protecting and Enhancing Comprehensives 81
Matting 82
Mounting 83
Preparing Dummies 83
Scoring 84
Folding 84
Making Package Dummies 85
Making Folder Dummies 86
Making Booklet Dummies 87

CHAPTER 5 ADVANCED TECHNIQUES IN THE GRAPHIC DESIGN PROCESS 91

Producing Advanced Comprehensives 91
Photocopies 91
Photostats 93
Photoliths 94
Color Keys 94
Enhancing and Protecting Comprehensives 96
Varnishing 96
Cellophane Wrapping 97
Acetate Wrapping 99
Simulating Printing Techniques 101
Embossing 101
3M Image 'N Transfer System 104
Rub-downs and the Chromatec Process 108
Silk Screens 110
Creating Special Effects 112
Photographic Materials 112
Creating a Grid 113

CHAPTER 6 PRESENTING DESIGNS 115

Choosing the Pieces to Present 116
Preparing for Presentation 116
Two-Dimensional Pieces 117
Three-Dimensional Pieces 117
Sketches and Layouts 117
Determining the Format 119
Choosing a Presentation Vehicle 119
Multi-ring Presentation Books 120
Multi-ring Presentation Cases 120
Rigid Presentation Cases 120
Advantages and Disadvantages 121

CHAPTER 7 DEMONSTRATIONS 123

Demonstration One: Package Design Assignment 123
Step One: Research and Analysis 124
Step Two: Thumbnails 124
Step Three: Rough Layouts 125
Step Four: Finished Layouts 126
Step Five: Comprehensive Layout 129
Step Six: Advanced Comprehensive 130
Demonstration Two: Annual Report Assignment 135
Step One: Research and Analysis 135
Step Two: Thumbnails 136
Step Three: Rough Layouts 137
Step Four: Finished Layouts 139
Step Five: Comprehensive Layout 141
Step Six: Advanced Comprehensive 141
Demonstration Three: Magazine Advertisement 145
Step One: Research and Analysis 145
Step Two: Thumbnails 145
Step Three: Rough Layouts 146
Step Four: Finished Layouts 147
Step Five: Comprehensive Layout 149
Step Six: Advanced Comprehensive 149

CHAPTER 8 STUDENT PORTFOLIO 153

Glossary 169
Bibliography 183
Trademark Information 185
Index 187

Introduction

The Complete Book of Comprehensives is a text for students and professionals and a reference for anyone in the graphic design field. People working in graphic arts production, such as suppliers, printers, and typographers, as well as those in the business end of the design field will gain a better understanding of how ideas are transformed from the mind to paper and how effective communication takes shape. Although it would be helpful if you have experience with or knowledge of basic design studio procedures, the information presented here can be understood by anyone.

In the graphic design and advertising fields, the development of an idea is a process that cannot be accomplished solely through prescribed methods or procedures. The process is unique for each designer, enabling the person to establish individuality, as evidenced in his or her work. This book is intended to help spark *your* ideas and provide the procedures that will allow you to develop them to their full potential.

In addition to procedures, you will find equipment, tools, materials, and techniques that are commonly used by graphic designers and art directors. In recent years there have been technological developments that have permitted the improvement of the quality of layouts and comprehensives. In fact, many layout and comprehensive rendering techniques and processes produce results that are better than those achieved in reproduction. Methods that have been and will be around for many years and ones that have been proved useful in recent years are presented.

The Complete Book of Comprehensives is practical. It explains and illustrates how and when each procedure, technique, piece of equipment, tool, and type of material is used. It anticipates the problems and situations graphic designers and art directors constantly face and gives exercises and demonstrations that provide useful solutions. In addition, a bibliography lists excellent books on related topics.

All the work in this book was rendered by students at art and design schools throughout the United States. Their work is used to illustrate how the procedures, techniques, tools, and materials will help you to produce successful layouts and comprehensives. The terminology used to describe materials, equipment, tools, techniques, and procedures is employed in art and design schools and by the profession.

Finally, it should be noted that new technology and skilled hands are tools for the execution of excellent visual ideas. There is no substitute for a well-rounded art and design education and work experience. It is the responsibility of professionals, students, educators, and suppliers to establish and maintain a dialogue that extends the boundaries of what we can achieve with todays' materials and procedures. It is the goal of this book to encourage this dialogue.

chapter 1

The Designing Process

In any given assignment the job of the graphic designer is to bring together many disparate elements—such as words and images—in a form that will communicate an intended message in a fresh and innovative manner. This is always a challenge because graphic design assignments usually must be completed in a very short amount of time, often within budgets that force a designer to be as creative in the use of time and materials as in the achievement of results.

The tight time constraints built into every deadline do not have to restrict your creativity—provided you have a solid plan for working. The most widely used work plan followed by graphic designers faced with a new assignment includes the following steps:

Research and analysis

Thumbnail sketches

Rough layouts

Finished layouts

Comprehensives

Advanced comprehensives

These steps, which are the foundation for this book, are followed religiously by professional graphic designers in order to save time and money by directing their efforts in a controlled manner. They serve a dual role for everyone involved in a project. For the designer, they help to outline and plan a work schedule that will lead him or her through various steps involving increasingly finished levels of work. For the client, these steps permit him or her an opportunity to offer input at various points along the way, thus avoiding any surprises at the end.

The first three steps can be categorized as roughs and the last three can be categorized as finishes. The roughs, which are the first steps in which the designer's thinking and raw energy become focused, will take many directions and forms and are intended to bring the assignment under control (Fig. 1.1). The finishes—the finished layout, the comprehensive layout and the advanced comprehensive—are generally the final creative steps undertaken by a designer when developing an idea, and for the most part these will determine whether or not a piece will be printed (Fig. 1.2). The last two terms are commonly referred to by designers and others as comprehensives, or "comps." Each represents a high level of finish and serves a similar purpose in the presentation of an idea.

The following chapters list the various equipment, tools, and materials and the basic techniques used to develop and actually produce your ideas. The text concludes with a discussion of various presentation styles, as well as a portfolio of student work.

Figure 1.1 ***(top).*** Rough sketches, or "roughs," are the designer's first scribbles on paper. They are intended to show the designer's thinking in a rough, loose form.

Figure 1.2 ***(bottom).*** "Finishes" are refined, meticulously executed sketches through which a designer develops an idea or ideas to a level of finish at which it can be evaluated to determine whether or not it works.

chapter 2

Basic Steps in the Graphic Design Process

Graphic design is a problem-solving process. What makes it different from other such processes is that problems are solved visually, on paper: A designer's thinking process is recorded as scribbled notes and sketches. Although there are no formulas for solving graphic design problems, there are steps necessary to follow to produce a successful solution.

STEPS IN SOLVING GRAPHIC DESIGN PROBLEMS

The number of steps needed to solve a graphic design problem varies according to each designer's level of talent and experience, but they generally include the following:

1. Research and analysis
2. Sketching thumbnails
3. Preparing rough layouts
4. Preparing finished layouts
5. Preparing comprehensives
6. Preparing advanced comprehensives

The latitude for experimentation within each of these steps is immeasurable. Because each designer's thinking and level of creativity is unlike anyone else's, an infinite variety of solutions can be found for each design problem.

The six steps were listed in order of their level of development and degree of finish. The first three steps, which are concerned with the development of the idea, generally end in what are called *roughs*, or *layouts*. Basically, these show the size and position of all elements, such as text, illustrations, and photographs. The last three steps, which are concerned with the completed idea and final execution of the design, generally end in *finishes*, or *comprehensives*; if the design is produced three-dimensionally, such as is done for booklets, packages, or brochures, it ends with what is called a *dummy*. These final products show accurately what the printed piece will look like, including all the details and fine points of the design. When executed properly, the last two steps bring the idea to a level that is barely distinguishable from the printed piece.

Because of time and budgetary limitations, it is not always feasible to adhere to all six steps to solve every design problem. They are intended as a guide. All the steps, however, are vital for a thorough development of an idea, and whenever possible they should be executed in order and treated with equal emphasis.

Each of the steps in the graphic design process has its own objectives, discussed in this chapter; requires the use of special equipment, tools, and materials, discussed in Chapter 3; and has basic and advanced techniques associated with it, discussed in Chapters 4 and 5, respectively. Throughout the process, it is important to become familiar with the purpose of each step, proficient in using the equipment, tools, and materials, and comfortable with the techniques. A discussion of each step in the graphic design process follows.

Research and Analysis

The design process generally begins with a briefing, a meeting of the client, designer, writer, photographers, and others involved with the project, during which the designer is told about the assignment and is given any available materials that may be pertinent to the particular project, such as competitive products (in packaging assignments), existing logotypes, corporate graphic standards, photographs or slides, descriptive copy or manuscript, and previously printed designs, which might help to describe the feeling or impact wanted from the finished product. The first step in any design project is to do some additional research and to carefully analyze the materials provided during the design briefing.

The research and analysis step is mandatory since it enables you to become familiar with the general category to which the particular project belongs (especially important for any packaging design assignment). It also enables you to familiarize yourself with the graphics used on competitive products, ensuring that your solution will be aimed at a particular target market (Fig. 2.1). It triggers your imagination and starts the flow of ideas and in many cases will bring

Figure 2.1. Careful analysis of the competition and the environment in which a product is displayed is an exercise that is always performed when beginning a mass-market oriented package design exploratory.

up possible solutions that you would not have considered otherwise.

When you begin any design assignment, there are many questions to ask yourself that will help you focus your thoughts and ideas in the direction of an appropriate design. For example, consider the following:

1. What is the message being communicated?
2. Who is the audience to be reached?
3. What form will the final product take?
4. Does the finished graphic look have to conform to a previously established corporate identification system?
5. How will the finished product be used?
6. How will the piece be printed?
7. Will the design include illustration, graphics, photography?
8. How long will it take to complete the project?
9. What are the budgetary limitations?

After the initial design briefing, carefully analyze the information you have collected. The size and printing limitations, assignment due date and budget, existing corporate graphic identification specifications, (usually a written statement of design requirements such as type, color and quality of artwork reproduction, and placement of logos or other identifying graphics, such as rules, bars, or bands) will help you determine what direction the project will take. For example, if a corporation's graphic look is a combination of a logo, a rule, an identifying typeface and/or color, these elements should probably be incorporated into your design.

Before you make any final decisions, however, do as much research as possible and explore as many sources as are available. A good place to begin your research is at the public library. There are general reference books and annuals that showcase award-winning photography and illustration and design solutions as well as specific graphic design periodicals, annuals, and books, such as *Print, Graphis, Art Direction, Communication Arts,* and many others that can be useful.

In addition, there are books and annuals issued by the American Institute of Graphic Arts (A.I.G.A.), the Art Directors Club, and various other professional organizations and clubs (Fig. 2.2). In fact, many designers begin a project by flipping through these books and periodicals for inspiration. Here the designer can get the creative ball rolling by exploring a number of sources. Some libraries also have picture reference files organized by subject (Fig. 2.3). For example, if you need a picture for a ballet poster assignment, you would look in the picture file under the heading Ballet or Dance. A comprehensive directory of sources for professional users of pictures can be found in *Picture Sources,* published by the Special Libraries Association in New York City. This directory lists special collections of prints and photographs on all subjects, available in the United States and Canada.

Specialized picture services, such as the Bettmann Archive, in New York City, have large collections of pictures on all subjects, which can be borrowed for a fee. There are also government agencies that supply pictures, many of which are listed in *Pictorial Resources in the Washington, D.C. Area,* published by the Library of Congress.

It is a good idea to begin building your own reference library of books and periodicals early in your design career. You will also find that book and magazine stores are another valuable reference source for ideas and designs. A good reference library will enhance any design studio in that it gives the designer ready access to visual material and ideas.

Figure 2.2. Many designers leaf through design periodicals and annuals for inspiration at the beginning of an assignment.

Figure 2.3. Examples of pieces gathered from picture reference files in two different categories at the New York Public Library. Picture reference material is categorized either by subject matter or by a person's last name.

In addition to libraries and bookstores, there are other resources for ideas. For package designers, supermarkets and drugstores are excellent reference sources for mass-market items from soft drinks to cosmetics. Specialty gourmet shops also offer many contemporary and international packages. By looking at shelves of items, you can evaluate the competition and use the information to help design an appropriate and successful piece. Annual report and other print design samples can be viewed at exhibitions, at printers, or by writing to corporations and asking for samples of their corporate and promotional pieces.

Sketching Thumbnails

After you have absorbed the information from the initial design briefing and your subsequent research, images will begin coming to mind that you might be able to use graphically. Quickly record all of the images that are in any way related to the assignment; you can sketch them on anything readily available, similarly to how you would jot down notes. The graphic design term used to describe these small, rough, uninhibited graphic notes or sketches is *thumbnails* (Fig. 2.4), so called because of their size, which is actually larger than a thumbnail but much smaller than the finished size. Because thumbnails are freehand drawings, intended to give only an impression of what the final solution will look like, they let you explore and develop many ideas easily. For example, if the intended design is to have a bold headline and flush-left text copy, the thumbnail would consist of a broad, thick freehand stroke for the headline and rules or loops aligned on the left for the text, all of which can be done in a few seconds. Further, you do not have to specify the actual typeface, but you should indicate the general height, weight, length, and position of the letters that you intend to use in the final piece with the techniques described in Chapter 4 for indicating text and display type. If the design is to have text, or *body* type, it is also necessary to estimate the space that the text will occupy and indicate it accordingly. When doing a thumbnail of a three-dimensional package or other object, you can render it two-dimensionally in perspective showing how the design works with the shape of the package, or flat showing the principal panel and adjacent sides.

The graphic design tools and materials used to make thumbnails are described in detail in Chapter 3. Thumbnails are usually rendered on tissue, layout, or bond paper, but virtually any paper will work. The materials used for black-and-white thumbnails are medium- to soft-graphite drawing pencils or a combination of broad- and fine-nib felt-tip markers. If color will be an element in the final design, as is often the case, experiment with it at this early stage. Colored pencils and fine-nib color markers are the most popular color media for thumbnails.

Through trial and error, a designer will come up with one or more design directions to develop further. Since the selection of a thumbnail to use for the next step in the graphic design process is crucial, you may wish to seek outside opinions such as from your instructor or art director at this stage. Furthermore, an idea that you lay aside might be perceived as worthy of development by someone else. By inviting others to look at your work, more options can be explored and successful solutions can be found.

Even after the final direction (or perhaps directions) has been decided, it is wise to save all of your thumbnails. Perhaps the initial presentation will not be adequate and you will need additional ideas, or maybe you will have a similar assignment in the future. Also, instructors, clients, or employers might want to see your thumbnails to get an idea of your thought process. When you have your thumbnails and have chosen the ones most appropriate for your assignment, you can proceed to the next step, rough layouts.

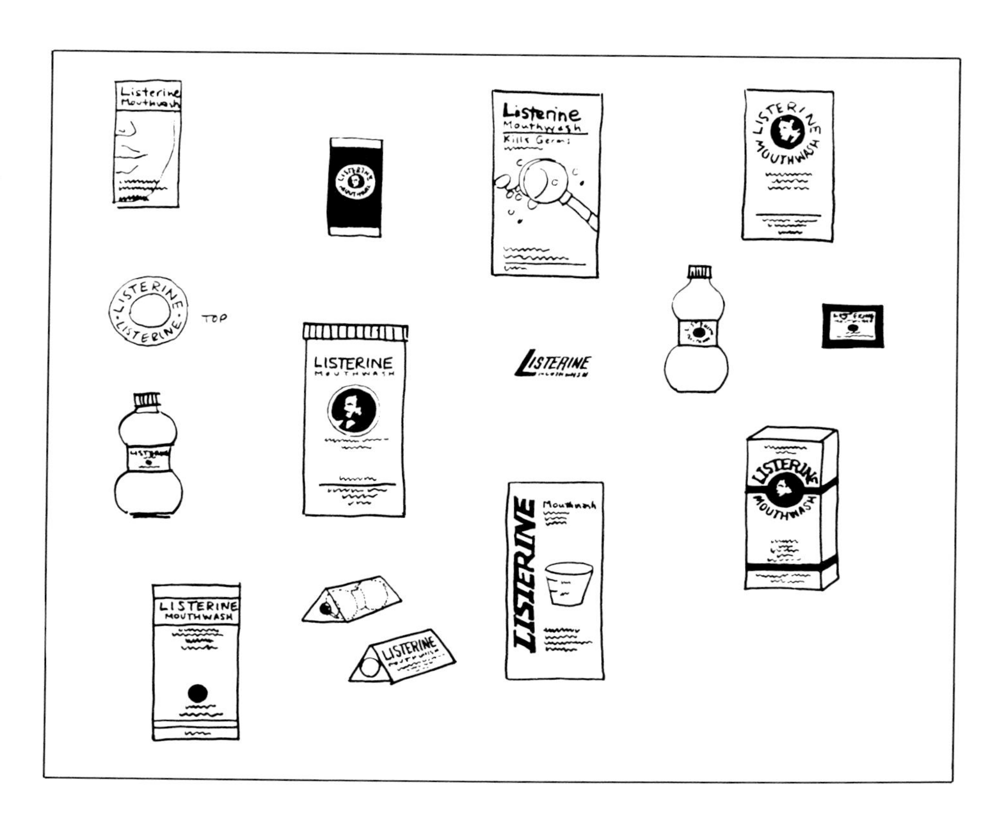

Figure 2.4. Thumbnail sketches that were generated in a matter of minutes using a variety of media.

Preparing Rough Layouts

After the most promising thumbnails have been selected, the next step in the graphic design process is to draw them at their reproduction size, which is the size the finished piece will be when it is printed (Fig. 2.5). The only exceptions to this are large posters, large cartons, displays, signs, billboards, and other items whose scale makes it impractical to do full-size sketches. In these cases, prepare the thumbnails proportionally smaller than they will be in the finished piece, using simple proportions such as one half or one fourth of actual size, in order to make translation to actual size easier to calculate.

These enlarged sketches are called

Figure 2.5. Thumbnail sketches are often translated to reproduction size by visual estimation.

rough layouts. A rough layout—or simply layout—is a sketch made using hand-rendering techniques (i.e., drawing freehand, or without the aid of T squares, triangles, templates or photographic reproduction) that shows a close approximation of the size, position, and spatial relationship of all the design elements (Fig. 2.6). The layout indicates the actual size, weight, and position of both *display,* or headline type and body, or text, type. It also shows the actual size and position of all pictorial elements, including photographs, illustrations, and graphics. When adding any pictorial elements, render them in solid blacks or gray tones as opposed to outlining them (which does not accurately reflect their weights). Keep in mind, however, that like thumbnails, layouts are rough sketches rendered in a loose, fast style. The display type is roughed in, that is, it is drawn quickly to simulate the character and spacing but not the letterforms. The body type is then *Greek*ed *in,* or its size, leading, and length is indicated by made-up, hand-drawn lettering.

The loose style of layouts makes it easy for designers to use them for experimenting, which indeed is one of their functions. After you have enlarged the thumbnail sketch, you can begin a trial-and-error process. By experimenting with shapes and forms and their spatial arrangement through enlarging, reducing, condensing or cropping the trim size around photographic images, rearranging, and editing, you transform your ideas into forms that can be viewed and evaluated by yourself and others. As you draw each new layout, evaluate it, revise it, and then place it under a new sheet of tracing paper so that you can trace the revised layout (Fig. 2.7). This revised layout is then further evaluated and revised and it becomes the underlying sheet for your next layout. By following this process, you can adjust the size and position of the design elements easily and quickly.

Figure 2.6. All elements of a rough layout should be drawn in a loose manner but their positions should be accurately indicated.

Figure 2.7. When making rough layout sketches, the underlying layout is used as a guide for maintaining or rearranging the various design elements.

As your rough layout approaches completion and the design becomes increasingly more refined, the typographical elements such as headlines and text should become more specific. Although you still do not have to decide upon specific typefaces, you should indicate the type height, weight, and position, the *leading* (space between lines of text), and the length and width of columns. Remember, the roughs have to be fine enough to be viewed and understood by others.

The tools and media used to make rough layouts, which are discussed in detail in Chapter 3, include the following:

- Tracing paper pad, which lets you trace photographs and other illustrative material easily and will not allow markers to bleed through. A disadvantage of tracing paper is that colors will appear muted because of the paper's grayish cast. To correct for this, back your design with a sheet of opaque bond paper when the tracing has been completed.
- Soft and medium graphite drawing pencils, gray markers, and technical pens for rendering black-and-white layouts.
- Felt-tip markers and colored pencils for rendering color layouts.
- Lucy (projector), or reducing/enlarging copying machine, for enlarging thumbnails to reproduction size.

Generally, you will select several directions from the thumbnails and develop each one. The main reason is that when some thumbnails are enlarged to the full size, they no longer work as a successful design. This may be because subtle relationships between design elements at a reduced size might not work when they are enlarged. Also many designs often tend to look good at a reduced size because there are fewer problems to resolve.

Another reason to develop several thumbnails is to have more than one design for comparison. For example, if you are redesigning an existing package, the client may want to have a choice of two design directions: a close-in design or one that is hardly different from the existing package except perhaps in color or typography, and a far-out design that is very different, with an innovative type, color, or illustration style, or a new shape and size. Often many degrees of far-out designs are shown to a client in order to arrive at a compromise between what the client and the designer would like to see.

Preparing Finished Layouts

After you have prepared rough layouts of several thumbnail sketches, analyze and edit them so that they represent sensible, strong design directions that seem to solve the assignment's objectives. For example, if a brochure for a retirement village contains type that is small and hard to read, then it will be a failure no matter how well it is designed. At this point, a decision should be made as to whether this problem can be resolved without sacrificing the design.

Now is a good time to have others help you. As mentioned, designers often do not realize the potential of a sketch, and the comments and criticisms of instructors, other designers, supervisors, and even clients will help you see your designs in a new way. When one or more rough designs have been chosen, the next step in the graphic design process is to render them as finished layouts.

A finished layout contains the same design elements—text, display type, graphics, photographs, illustrations, and so on—as roughs, but in a clearer and more refined state (Fig. 2.8). Actually, the distinction between a rough and finished layout is not always clear. Sometimes the rough is refined to the point that it is a loosely rendered finished layout, and sometimes the finished layout is actually a well-done rough. What distinguishes them is the media used to render them and their places in the graphic design process. The aim of the rough layout is to narrow down your selection of thumbnails and to enlarge some of them to reproduction size, whereas the aim of finished layouts is to narrow down your selection of roughs and bring them to a higher degree of refinement. Also, it is at the finished layout step that you get an idea of whether the design elements are compatible with one another. This is why you indicate the illustrations and photographs as masses or solid tones rather than outline shapes and represent text type by thin or thick loops or lines to show height, weight, leading, and column depth clearly. Headline type can be traced from a type book, or its individual characteristics can be created freehand.

How refined the finished layout is depends upon its purpose and to whom it will be presented. If the finished layout will be judged by an instructor, experienced designer, or someone else who is used to looking at layouts, it can be shown in a rougher state than if it is to be judged by a new client or someone unfamiliar with the design process. For some pieces, especially those with a limited budget, the finished layout is the last step before design approval and production. These layouts are necessarily more refined.

Regardless of how rough or refined your finished layout is, the level of finish of all of its elements should be the same. If you rendered the artwork in a light, loose manner, such as indicating the size and position of subject matter within an illustration or photograph using a broad marker or colored pencil, the typography should also have a light, loose feel. If the artwork is tightly rendered—for example a photocopy of an actual illustration or photograph—the type should be rendered to the same degree.

Figure 2.8. The finished layout should provide an accurate indication of the size and position of all artwork and typography. Photographs and illustrations should be indicated as tones rather than outlines. (Artwork: Steve Powell)

Finished layouts are usually rendered on *layout* and *visualizing paper,* which has a translucent, resilient surface, allowing you to trace designs without the aid of a light box. Tracing paper allows the accurate tracing of underlying artwork or type with all media, including markers, without the danger of bleed-through.

Bond paper, which is opaque and stiff, is also used in some cases, such as in making three-dimensional dummies for packages and brochures. For black-and-white finished layouts, markers, which are available in a wide selection of grays and blacks, and technical pens, which come in an assortment of nib shapes and widths, are often used. These are excellent for rendering type, and indicating halftone photographs and illustrations. For color layouts, markers, colored pencils, and technical pens are occasionally used, depending on the type of design. Technical pens are excellent for body text and small type (Fig. 2.9). Markers and colored pencils, which can produce a wide range of different effects when used together, are the best media to use to render in color: Broad-nib markers are good for rendering large areas of color, and fine-nib markers for filling in the details; colored pencils are used to shade shapes and add depth and highlights. Spirit- and water-based markers will not bleed into one another, so they can be used at the butting edges and overlapping areas of color; and the color in water-based markers will "bead up" when applied over colored pencils that have a high wax content, creating a smooth, even surface over broad background areas.

The way in which you present a finished layout depends on the type of paper on which you made your layout and what form the finished product will take; for example, see Figure 2.10. If you used layout or visualizing paper, back the layout with heavyweight bond paper with spray adhesive or staples so that you get an accurate reading of the colors. If you used bond paper, which is opaque itself, you probably will not need a backing. In either case, put a mat window over all the flat pieces; if the design is a three-dimensional brochure, book or package, it should be adhered onto a blank dummy to make it ready for presentation (Fig. 2.11).

Preparing Comprehensives

The next step in the graphic design process is preparing *comprehensives* (or "comps," as they are commonly called), also referred to as comprehensive layouts. Comprehensives are executed after questions concerning the design have been resolved in the finished layout and all earlier stages of the design process. The research, analysis, and experimentation in these earlier stages will usually allow you to produce a design solution that requires few or no changes. Slight refinements that do have to be made can be done fairly easily.

Figure 2.9. Body copy being rendered with a technical pen on a finished layout.

Figure 2.10. This finished layout was rendered with color markers on layout paper and reproduced on a color-copying machine for presentation purposes. (Artwork: Jay Shmulewitz)

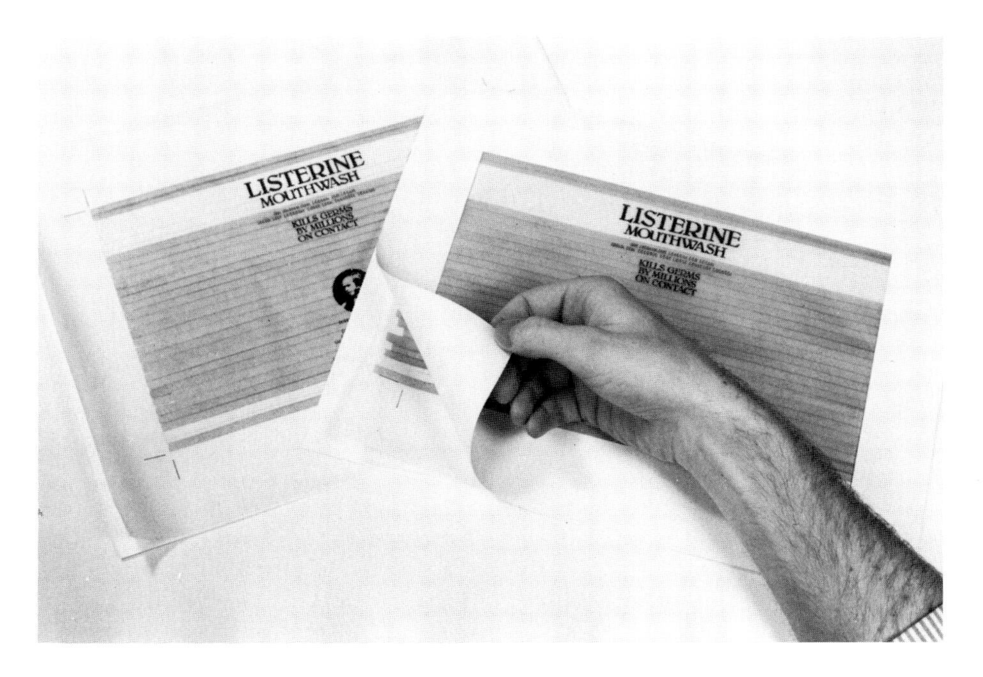

Figure 2.11. Two different methods used to mount finished layouts for presentation: the right example shows a window mat placed over a bond paper-backed layout; the left example shows a layout that has been stapled to bond paper to provide an opaque backing.

Comprehensives differ from finished layouts in the tools and media used in their production as well as in their level of finish. The main purpose of comprehensives is to simulate a reproduced piece, that is to create a prototype, using only traditional hand-rendering techniques and materials, including many of those used in the preparation of finished layouts, such as markers, pencils, and pens but also using high-quality rag layout paper as well as the actual paper stock to be used in printing (Fig. 2.12). Other tools, such as an airbrush or LetraJet might also be used to create tones, gradations and artwork. Since comprehensives are done solely with hand-rendering techniques, the main requirements for producing beautiful comprehensive layouts are a good imagination, a knowledge of using the right tool or medium to accomplish the desired effect, and competent rendering skills (Fig. 2.13).

Comprehensives enable instructors, clients, designers, and consumers to

review and evaluate a design idea as if it were a final, printed piece and to make last-minute changes. They also serve as a way to evaluate alternative designs quickly and make a final decision as to which one will be printed. Therefore, all design features, both text and graphic, must be shown as accurately as possible, including their styles, sizes, and positions. Techniques used to render the text and graphics on comprehensives include hand-tracing type from a typeface book, adhering actual photographs or hand-rendered illustrations to a layout, or blind-embossing type or artwork on actual printing paper. Comprehensives can also serve as a guide to printers for layout, or position of design elements, folding methods, and color breaks, that is, showing what areas will be in what color.

Comprehensives can be rendered on different papers ranging from layout or bond paper to the actual colored printing papers that will be used for the final piece. Comprehensives of flat or two-dimensional pieces are often produced on a stiff board such as illustration board, on which liquid media such as acrylics, designer's colors, and dyes can be used without resulting in the board's buckling or warping. Also, window mats and acetate can be placed over illustration board without its having to be backed with a board for support, as lighter-weight papers must be.

Figure 2.12 *(top).* The comprehensive layout simulates a reproduced piece solely with hand-rendering techniques. This example employs three different medias: type rendered with India ink, figures rendered with gouache, and wrench rendered with cut paper. Illustration board is used as the base for all medias. (Artwork: Scott Santoro)

Figure 2.13 *(bottom).* A comprehensive layout's success depends greatly upon the designer's rendering skills. Not only must the idea be effectively communicated, but it must also be well executed. (Artwork: Steve Powell)

When preparing three-dimensional comprehensives, such as for booklets or packages, you can render the comprehensive on lightweight paper or acetate, tape or cement it into the shape you want, and then assemble it for presentation.

The media most often used for rendering comprehensives are designer's colors, dyes, inks, acrylics, and color papers, films, and cellophane (Fig. 2.14). In some cases, photographs and photocopies are used to produce effects that cannot be done as fast or as effectively by hand. For example, if a project is on a tight schedule, or the subject matter needs to be shown literally, a photocopy or an actual photograph ("clipped" from a printed piece), which closely represents the photograph's subject matter is usually acceptable. Also, type, line figures, artwork, rules, textures, and patterns can be rendered (in black) on the actual printing stock hand-fed through a photocopying machine.

There are many methods you can use to render comprehensives, differing in level of difficulty and degree of finish. The one you choose should approximate as closely as possible the desired look of the reproduced design. For example, if you are designing a soft drink container, which will have the graphics on a metal cylinder, render the comprehensive on a medium (such as Mylar or foil) that looks like metal, or render it on acetate that you then wrap around a bare aluminum cylinder. If you are trying to simulate a matte varnish effect on a highly glossy paper, render the comprehensive directly onto a high-gloss printing paper. Once your comprehensive has been rendered, reviewed, and finalized for presentation, you can go the last step in the design process—preparing advanced comprehensives.

Preparing Advanced Comprehensives

An advanced comprehensive (also called a "super comp"), serves the same purpose as a comprehensive in terms of presenting a design in a highly finished manner, but it differs greatly in the materials and methods it uses. Whereas comprehensives are done using only hand-rendering techniques, advanced comprehensives use precise photographic techniques such as photostats and photolith films (also called Kodaliths), 3M-I.N.T.'s, rub-down transfers, photographic prints, typeset type, and other methods that produce results that are indistinguishable from those used on the printed piece (Fig. 2.15). *Line* and *halftone* copy, such as type and photographs, are often used to prepare advanced comprehensives to ensure accurate simulation of a printed piece. In addition, almost any special effects that can be achieved through printing methods, such as embossing and spot varnishing, can also be produced with advanced comprehensive techniques. Advanced comprehensives of three-dimensional packages, booklets, or folders are created directly on or adhered to dummies or whatever final form the finished piece will take. Chapters 4 and 5 discuss in detail the techniques used to prepare

Figure 2.14. A comprehensive layout rendered with acrylics that have been thinned with water and applied by brush.

Figure 2.15. Advanced comprehensives are rarely distinguishable from printed pieces because of the photographic techniques and equipment used in their production. (Artwork: Bruce Hanke)

advanced comprehensives and to create special effects.

Although there are times when it is desirable to render both a comprehensive and an advanced comprehensive, for the most part preparing both is repetitive. Either one can be used as a precise indication of what the printed piece will look like. The choice between whether to render a comprehensive or an advanced comprehensive to simulate a designed piece would seem to be an easy one. Since an expertly executed advanced comprehensive is indistinguishable from a printed piece, it might seem that all design ideas should be rendered in this manner. Other factors, such as time, budget, availability of equipment and materials, and the audience, however, often affect the choice of how to execute and develop a design idea. If little time or money is available to develop an idea, which is often the case, your client or instructor may not want you to prepare an advanced comprehensive. Also, unless you own or have access to the necessary expensive photographic equipment and materials, advanced comprehensives cannot be produced. In addition, many art school instructors insist that students hand-render all elements of a piece so as to develop their skills in rendering type and artwork and handling tools and materials.

The audience to whom you are presenting the design can also determine whether you produce a comprehensive or an advanced comprehensive. If the design is to be evaluated by people who know how to interpret a design, such as instructors or experienced designers in a studio or agency, it can be rendered as a comprehensive. If it is to be evaluated by someone who is unfamiliar with graphic design media and procedures, such as a client or sales or marketing representative, it should be rendered as an advanced comprehensive, which has a more finished look.

When an idea has been thoroughly developed and refined in the layout stages, it is easier to produce the final, advanced comprehensive. Many of the problems with the design will have been worked out, so only those that have not been resolved, such as specifying type, taking photographs, or rendering illustration artwork, will have to be dealt with at this last stage.

The importance of preparing well-executed advanced comprehensives cannot be stressed enough. If your idea is well executed, it can sell itself. If your idea is good but poorly executed, selling it can be difficult, especially if it is being presented to someone unfamiliar with the graphic design process or not capable of interpreting a design idea. Also, when more than one solution to the same assignment is presented and one is better executed than the others, the latter may be chosen simply because of the quality of its execution.

Although some people see advanced comprehensives as expensive to produce, if you have the equipment and material they are actually inexpensive in comparison to the printing costs of producing the final piece. In fact, they can save money, since they provide an easy, comparatively inexpensive way to experiment with an idea and avoid unwanted surprises in the finished, printed piece, as well as present alternative solutions to a design problem. Elements can be added to or removed from the advanced comprehensive. Also, options such as color and paper can be explored easily.

Unlike the pieces produced in the other steps of the graphic design process, advanced comprehensives are used for more than working out and presenting a design idea. One use is for consumer focus and test-marketing groups, which test new or changed products. (In these cases, advanced comprehensives are placed on supermarket shelves or in the hands of "consumers" to give the impression that the comprehensive is really the printed product.) Another use is in print and television advertising, where crisp, clean results are needed. Higher-quality results can be achieved with advanced comprehensives, which are produced one at a time by photographic methods, than with printed pieces, which are mass-produced. Furthermore, designs often need to be simplified and slightly altered when used in magazine and television advertising so that they appear larger, brighter, and bolder and to ensure fast recognition and readability, all of which can be done with advanced comprehensives. Although most comprehensives and advanced comprehensives are created as previews or simulations of what an entire finished piece will look like and thus a part of the final solution to a design problem, some are created as just one part of the whole comprehensive as a three-dimensional graphic element, or prop. These comprehensives can be rendered by hand or machine and later photographed and incorporated into the design.

Props are used in advertising and promotion to create an illusion, sensationalize an idea, attract attention, make an idea memorable—in other words, to do something to sell the product (Fig. 2.16). As an example, consider an assignment to produce a catalog cover for outdoor, "sportsmans" clothing. A solution using the technique of creating a prop was to use a duck decoy—an idea that is clearly out of context but that creates a memorable impression (Fig. 2.17). The designer paid careful attention to details so as to make the visual concept believable: The decoy was chosen for its lifelike quality and inherent personality and was outfitted with custom-tailored outdoor garments actually being promoted in the catalog. Whether the overall design solution would be successful would depend on the environment in which the decoy would be placed and how it would be photographed. From sketches made during the layout stages, the designer saw that two other decoys should be added; they were floated in the water in the background and blurred slightly so as to be indistinguishable from real ducks and so they would not distract from the "comprehensive," or prop, in the foreground.

Figure 2.16 Large-scale comprehensives using graphic-design techniques that are often used in television commercials for extra emphasis. (Courtesy of Dale Mallie, Mallie Studio, Inc.)

Another way in which comprehensives are used in advertising is as product packages created as oversized props for a larger-than-life, humorous effect. For example, giant dancing deodorant cans used as props in a particular television commercial were nothing more than oversized comprehensives. These oversized comprehensives are produced using basically the same materials and media as those used for conventional advanced comprehensives, but are executed with equipment specifically made for large-scale items.

Sometimes ideas are executed in three-dimensional form using found objects, such as wrapping a glass bottle with furry material as a solution to a package design assignment for a hair tonic. When the objects are arranged in a logical (or illogical) manner and then photographed, they can produce an excellent solution to a design problem. The objects can be used in their found state or can be altered with paint or other media to custom-tailor them to your needs. As shown in Figure 2.18, when properly incorporated into an effective layout using photographic means, compre-

Figure 2.17. Duck decoy used as a prop to advertise sportsman's clothing. (Artwork: author)

Figure 2.18. When props or comprehensives are photographed and incorporated into a strong layout, they can be very effective as design solutions. These solutions to assignments from instructors at the Pratt Institute in New York City include, clockwise from left to right, package for Hungarian Hair Tonic (a fictitious product), a prop for a catalog cover, and two separate solutions to be incorporated into a full-page magazine layout for Brink's Inc. (Artwork: Paul Graboff, author)

hensives or props of found objects can provide solutions that are as successful as those using more conventional media and materials.

A final and important use for both comprehensives and advanced comprehensives is as samples in your portfolio. They can be shown along with the printed piece, but they can also be shown alone. For example, ideas that have been executed but not shown to a client or that have been rejected by a client can be included, as long as they are well executed and show a strong concept. Also, comprehensives and advanced comprehensives of design solutions that were liked by a client but not printed because a project was canceled or postponed can be used for presentations and reference.

In general, layouts, comprehensives, and advanced comprehensives are important parts of the graphic design process and are necessary in order to go from a design idea to a finished piece. The chapters that follow present the equipment, tools, and materials you will need to produce layouts and comprehensives, the basic and advanced techniques you will need to learn, ideas about how to present your designs, demonstrations showing the use of the six steps discussed in this chapter, and a sample portfolio.

chapter 3

Equipment, Tools, and Materials

As a graphic design student and professional, you will undoubtedly receive a wide assortment of assignments. To carry out these assignments successfully, you will need special equipment, tools, and materials. This chapter introduces the equipment, tools, and materials, explains what they are, tells you how and when to use them, and describes their functions in the preparation of layouts, comps, and advanced comps. Later on, Chapter 6 explains and illustrates ways in which they can be used to produce special printing and finishing effects.

EQUIPMENT

Whether you set up your own design studio, work in someone else's, or use part of a room as your workplace, you will be able to produce better solutions to your graphic design problems if you are comfortable, have your tools and materials conveniently arranged, and have the necessary equipment, as shown in Figure 3.1. The following is a list of the furniture and equipment you will need to own or have access to.

Chairs and Stools

When you are working on a graphic design assignment, you sit for long periods of time, so a comfortable chair or stool is a necessity. You can choose among drafting chairs (Fig. 3.1), secretarial chairs, and artist's stools. Drafting and secretarial chairs are generally more expensive, but they have features that warrant their higher cost, such as swivel bases, pneumatic or manual height and backrest adjustments, and casters for mobility. If you decide to use an artist's stool you can make it more

Figure 3.1. A practical, comfortable, well-organized space complete with all necessary furniture and supplies, situated for easy access, creates a productive studio environment and an enjoyable place in which to work.

comfortable with extras such as a cushion, backrest, variable height adjustment, and swivel base.

Drawing Boards

A drawing board is a portable wooden board that can turn any desk, table, or countertop into a ruling and drawing surface (Fig. 3.2) on which you can produce accurately drawn layouts. It can be used flat or propped up with blocks to create a comfortable working position; some have a built-in tilting device. Generally, drawing boards have two metal straight edges along each side to enable you to draw more accurately with a T square; those that do not can be easily fitted with a True-Edge (see page 25). All types of drawing boards are available from 16 by 20 inches to 30 by 40 inches.

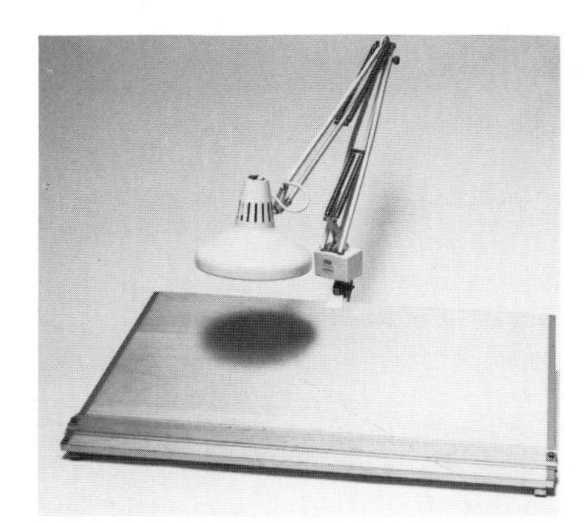

Figure 3.2. Drawing board equipped with a sliding parallel rule and a Luxo combination fluorescent/incandescent lamp. This portable set-up is inexpensive and can be placed onto a flat desk or countertop.

Drawing Tables

All drawing tables consist of a drawing board mounted on a base and feature tilt and height adjustments, but they differ greatly in the materials they are made of and their capabilities. There are, for example, models that fold up for easy storage as well as the traditional all-wooden pedestal tables, steel central pedestal tables, and tables that have either pneumatic or hydraulic height adjustments.

The drawing board part of drawing tables is made in a wide range of sizes and materials, but a 30-by-40-inch wooden or plastic top will accommodate large artwork and a number of supplies. *Note:* If the table top is not equipped with a true metal edge (the majority is not), a True-Edge, parallel rule or Glideliner will need to be attached to the board.

Since you will be cutting and using messy media, both of which can quickly ruin the surface of most drawing boards, you should protect yours. An inexpensive way to protect a drawing board is by taping a large piece of *chipboard* or *illustration board* over the entire work surface. Usually grayish in color, chipboard is an inexpensive, rigid, all-purpose board, usually used for backing artwork or constructing models. Illustration board is more expensive but offers a smooth, clean surface on which to work. A more permanent method is to apply a self-sealing vinyl cover that heals, or closes, cuts and holes instantly, leaving an unbroken surface. The vinyl can be cut to any board size and can be attached easily with two-coat rubber cement.

Taborets

A *taboret* is a storage unit for tools and supplies. It can be made of wood, steel, or plastic and is usually mounted on casters for easy movement and positioning. A taboret helps you organize and store your tools and materials and provides fast, easy access to them. Additionally, it enables you to reduce the volume of material that normally accumulates on the tops of drawing tables.

Rotating Trays

Rotating trays are compact, compartmentalized tabletop trays that can be used to hold brushes, pencils, markers, pens, knives, erasers, pushpins, and other commonly used graphic design studio materials (Fig. 3.3). The trays' small size makes them suitable for the top of your drawing table.

Lamps

Since graphic design work requires extreme accuracy and precision, you will need a good lighting source over your work area. The best light source is an adjustable-arm lamp that is fluorescent, incandescent, or a combination of both. It should have a long, easily adjustable arm that can stretch to illuminate any part of the work surface as well as remain fixed in one position.

Three lamps commonly used by graphic designers are the twin-tube fluorescent, the single bulb incandescent, and the combination fluorescent/incandescent. The twin-tube fluorescent lamp, with two 15-watt fluorescent tubes housed in a rectangular metal shade, will illuminate the largest area and is coolest to work under. Fluorescent tubes should be purchased in both cool white and daylight in order to achieve color-balanced light. The single-bulb incandescent lamp is popular with students and beginners because it is inexpensive, but it is very hot to work under and does not provide color-balanced light. The most popular lamp for general graphic design work is the Luxo combination fluorescent/incandescent. It has a 4½-foot adjustable arm with a 9½-inch-diameter metal shade that houses a 22-watt circular fluorescent tube and a 60-watt incandescent bulb for color-balanced lighting.

Figure 3.3. Rotating tray for compact storage of frequently used tools and supplies.

Light Boxes

A *light box* is a frame or open box that houses two or more fluorescent bulbs and is covered with a smooth piece of frosted glass or translucent plastic (Fig. 3.4). It is primarily used for tracing layouts or comprehensives, viewing transparencies and slides.

You can purchase commercially produced light boxes in many sizes and variations, but they are usually quite expensive. You might prefer to put together an inexpensive substitute from materials easily found in your home or studio. One way is to lay a piece of plastic over a 6-inch-high stack of books, an empty drawer, or an open crate and place a light source underneath. Another simpler but less stable way is to lay a piece of translucent plastic cover over an inverted rectangular fluorescent lamp.

Loupe

A *loupe* is a small 8X magnifier that is used to study negatives, contact sheet photographs, photostats, slides, and printed pieces to evaluate the clarity of the image (Fig. 3.5). Its clear

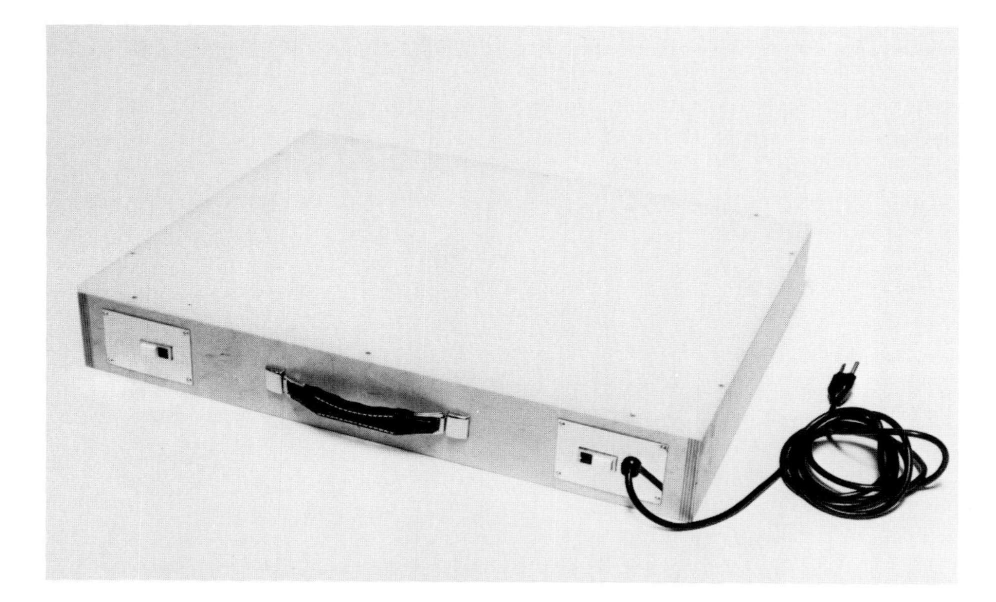

Figure 3.4 *(left).* Portable light box.

Figure 3.5 *(above).* Agfa Loupe.

plastic base holds the magnifier in focus while providing a distortion-free image. A loupe is invaluable when reviewing and editing slides and contact sheets before making photographic prints.

Hair Dryers

Hand-held electric hair dryers are used to expedite the drying time required for many materials and processes. They can be used, for example, to dry inks on drafting Mylar or acetate, enabling you to work fast without smearing. When adjusted to a cooler setting, hair dryers can be used to dry photostats, films, and photosensitive products. All of these processes and materials are discussed in Chapters 4 and 5.

Lucy

A *Lucy* (short for Lacey-Luci) is a projector used in the graphic design studio to enlarge or reduce opaque artwork, type, layouts, photographs, or transparencies (Fig. 3.6). Most models have an illuminated copy board and a telescoping lens unit that projects the image from the original

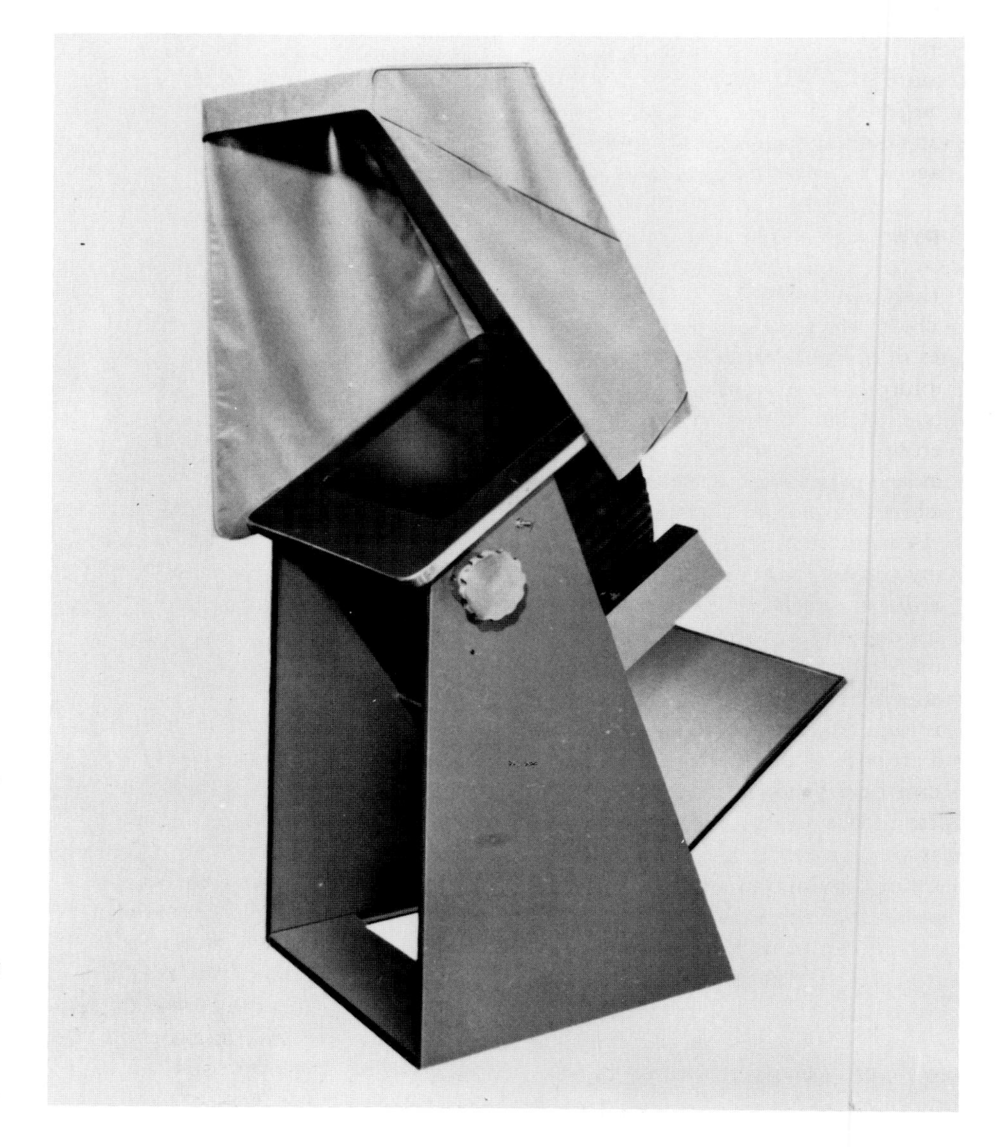

Figure 3.6. The Goodkin Model 5B Viewer, one example of a Lucy.

copy through a ground-glass surface onto a transparent material, such as tracing or visualizing paper, on which the image is traced. Tracing onto any opaque material, such as illustration board, is impossible with this projector.

Exposure Units

Exposure units are compact, portable machines used to expose photolith films through Color Key, Image 'N Transfer, Chromatec, and other light-sensitive products. They have ultraviolet fluorescent tubes that provide an even distribution of light and need no warm-up time. A built-in timer automatically shuts off the lights after the exposure is completed. Some exposure units have a vacuum-glass feature that presses the photolith film tightly against the photosensitive product so that the image is transferred accurately.

Copying Machines

In the graphic design studio, copying machines are used for making duplicates of color and black-and-white graphic material such as tissue layouts, comprehensives, and pasteups; reference materials; and type samples. Standard paper sheet sizes for copying machines are 8½ by 11 inches, 8½ by 14 inches, and 11 by 17 inches. Many copying machines can enlarge or reduce materials, or copy them on two sides as well as collate. In recent years, this machine has become an invaluable tool in graphic design studios for enlarging and reducing type and artwork to be used as guides for tracing or for showing the position of halftone artwork in a layout, comprehensive, or mechanical.

Color copying has been gaining in popularity and especially suited machines can be used to enlarge or reduce color photographs, printed pieces, and layouts and reproduce black-and-white pieces in one or more colors. Most color copying machines also have an attachment that will enlarge a color slide to fill a standard 8½-by-11-inch sheet of paper.

Photostat Machines

Graphic designers use photostat machines to reproduce black-and-white line copy onto photographic paper and photolith film and to create line or dot conversions from continuous-tone copy—such as photographs or illustrations (Fig. 3.7); these are then called continuous-tone prints or veloxes. These machines vary in their capabilities, but most provide sharp, clean enlargements up to 200 percent and reductions down to 50 percent of the original in a single shot.

There are two types of photostat systems: the two-step negative/positive system and the one-step direct positive system. In the two-step negative/positive system, first a negative is made from the original black-and-white copy. The negative is then used to make a positive, and you end up with a black-and-white photostat or copy that is the same as the original. To avoid confusion when you order photostats from an outside source or supplier that uses the two-step system, refer to the *negative* as the first print and the *positive* as the second print. A photostat can also be made with the image facing the direction opposite that of the original; this is called a flopped photostat (Fig. 3.8).

In the one-step direct positive system, a positive image (one that is the same as the original) is made directly from the original in one shot. This

Figure 3.7. Agfa-Gevaert "Repromaster 2001" photostat machine.

Figure 3.8. *a.* Original line art, *b.* negative print, *c.* positive print, *d.* flopped positive print.

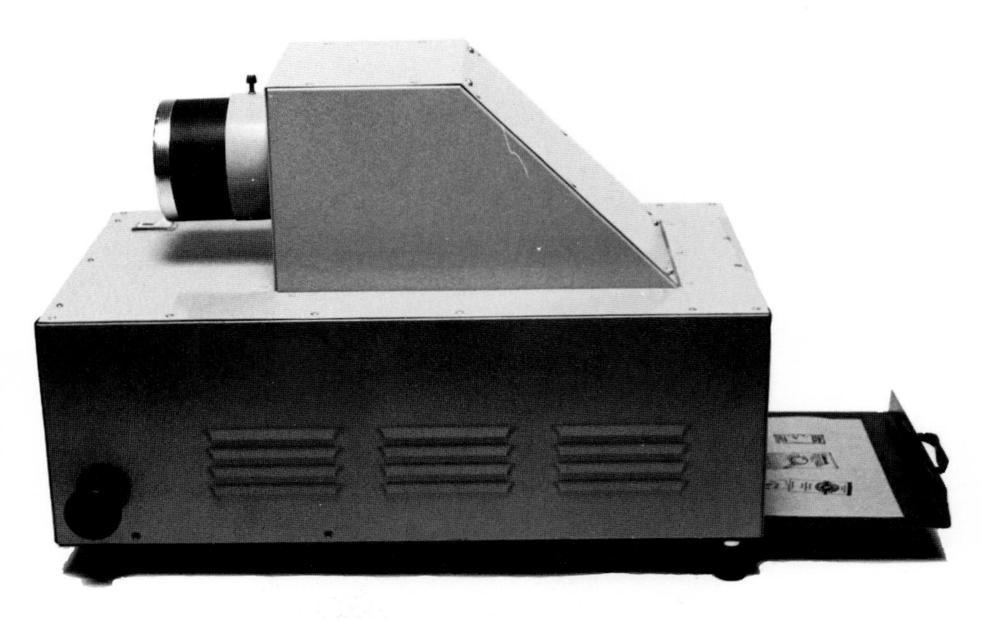

Figure 3.9. Opaque projector with slide-out tray that holds the original art.

system is also capable of making a direct negative (or reversal) in one shot. This means that everything that was black on the original will now be white and vice versa. When ordering photostats from this system, specify either a direct positive or direct negative to be made of the original.

Opaque Projectors

An opaque projector is used to project any opaque artwork, type, layouts, photographs, or visual/reference material as large, bright, undistorted images (Fig. 3.9). Any line or halftone copy that is flat or in a book, magazine, or newspaper can be enlarged with edge-to-edge sharpness.

Some opaque projectors are made for desk or countertop use, and others can be attached to drawing boards and projected directly onto the working surface. Opaque projectors are an excellent tool to use in preparing comprehensives because an image can be projected directly onto any opaque material, such as illustration board or printing paper.

Condensing and Expanding Lens

A condensing and expanding lens, called an *anamorphic lens* (Fig. 3.10), is used in the graphic design studio to condense or expand type on a photostat machine, eliminating the need to render these special effects by hand. Its only limitation is that for good results, the height of the capital letters of the type has to be one quarter inch or less. If you want to use larger type, you must first reduce it to this size.

To use a condensing and expanding lens, place the type to be enlarged or reduced in the center of the copyboard of a photostat machine, perpendicular to the lights. Next, place the "curved" lens on top of the type with either its highest point over the type (for condensing) or its lowest point over the type (for expanding, or squatting) and align

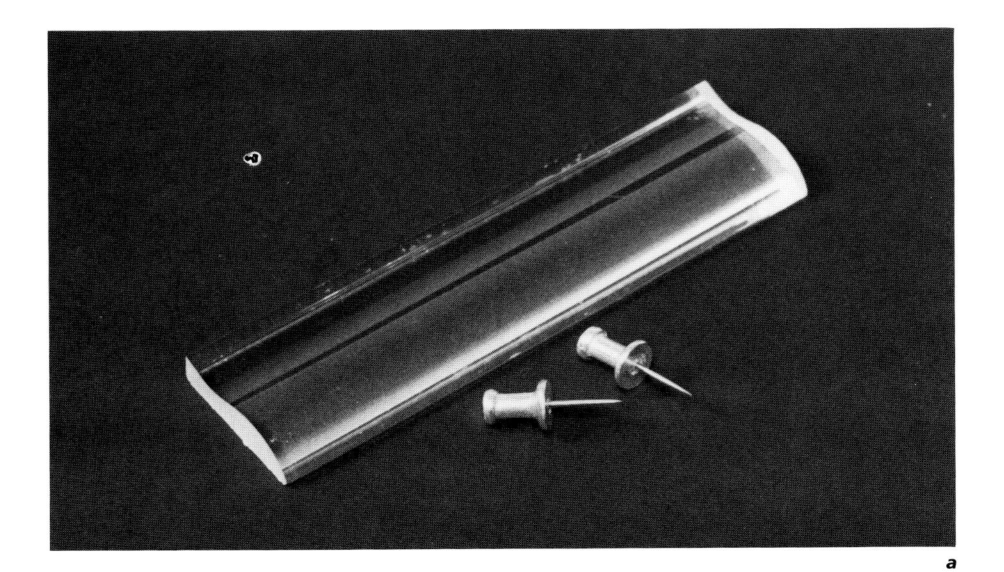

a

LISTERINE

LISTERINE

LISTERINE

b

Figure 3.10. *a.* Anamorphic lenses, used to condense and expand type; *b.* type that has been altered using the anamorphic lens on a photostat machine, showing from top to bottom, original, condensed, and expanded versions.

the lens so its top and bottom are parallel with the baseline, or bottom edge, of the type. Then make a normal exposure with the photostat machine. You can repeat this process several times to achieve a very elongated or squat effect, but keep in mind that the image will become more and more distorted with each photostat and you may have to redraw it by hand to clean it up.

TOOLS

Once you have the necessary equipment, you need to begin thinking about and collecting the tools and materials you will be using as a graphic designer. This section will describe the tools, which can be thought of as items that you use over and over, such as scissors and rulers. It will then describe materials and media, which are supplies that will be used up and need to be replenished, such as paper and ink.

Rules and Scales

The first category of tools includes rules and scales. In general, these are used for measuring, drawing straight lines—called ruling—and making straight cuts.

Rulers

Steel or hard-tempered aluminum rulers with etched-in calibrations are used as measuring devices and straightedges (see this chapter under Straightedges) for cutting (Fig. 3.11). They are available in lengths ranging from 12 to 48 inches and in scales measuring inches, millimeters, *pica*s, and agates. The 24- and 36-inch rulers are long enough to measure or trim most design pieces without being too cumbersome.

Parallel Ruling Straightedges

There are two types of parallel ruling *straightedge*s. The first type, called a parallel rule, uses a clear plastic-laminate or stainless steel blade (or rule), which is attached by a cord and ball-bearing pulleys to the drafting table top. The second type, called a Glide-Liner, uses a metal blade that is attached to a ball-bearing device fixed onto either edge of the top and side of the drafting table. The ball-bearing device is hinged, allowing the blade to swing away from the artwork.

The best feature that parallel rules offer is length. They can be purchased in lengths up to 60 inches. For most graphic design projects the metal-edged blade will function better for cutting than the plastic-edged one. The main drawback to using parallel rules is that the cord or wire that allows it to move freely will loosen and/or break. Also, they prove cumbersome when slipping tracing pads underneath to trace layouts and when they are moved over artwork.

Glide liners combine the advantages of a T square and a parallel rule. It stays in position, allowing you to work with both hands free (see this chapter under T Squares). Its hinged blade swings away without disturbing artwork, which also permits tracing pads to be placed under the blade for easy tracing of layouts.

Straightedges

Heavy stainless-steel straightedges (without calibrations) are used for ruling and cutting mat boards, cardboard, paper, fabrics, glass, and other materials. Most straightedges have one square edge and one beveled edge. They range in lengths from 18 to 72 inches.

True-Edge

A True-Edge is a metal device that clamps onto a drawing board or table to give it a straight and accurate surface or edge for ruling with a

Figure 3.11. Rulers *(left to right):* Circular proportional scale, Schaedler Precision Rules, 30°/60° plastic triangle, steel T square, aluminum ruler, type gauge, and 45° plastic triangle.

T square. A True-Edge will turn any non-smooth or crooked table top into an accurate drafting surface. It is available in lengths that match standard drawing-board sizes. True-Edges have built-in grooves, enabling you to slide a T square up and down with little friction. This is an important feature because of the constant shifting required in rendering most layouts.

T Squares

A *T square* consists of a thick cross-piece or head attached at a right angle to a long, thin blade and is used to draw and rule horizontally. Since you will use a T square in every aspect of your design work from rendering lettering to preparing mechanicals, it is important that this tool be of the finest quality. T squares are available in many materials, including wood/acrylic, aluminum/plastic, and stainless steel. Stainless-steel T squares are the most expensive but are best suited for all-around use in the graphic design studio. Their virtually indestructible all-steel construction lets you use them as a guide for cutting, and the beveled edge at the underside of the blade minimizes ink smearing. Only stainless-steel T squares should be used for cutting: all others can become damaged and thus useless.

T-square blades are available in lengths of 18, 24, 30, 36, 42, and 48 inches to match standard drawing-board sizes. The most popular lengths are 24 and 30 inches, which will accommodate most designers' needs.

Triangles

Triangles are acrylic or metal triangular tools that are used with a T square or parallel rule to rule or cut vertically (Fig. 3.11) or on an angle, depending on its configuration. Standard configurations are 30/60 degrees and 45 degrees. The 30/60-degree triangle is the most useful because its one tall edge permits the drawing of longer vertical rules on a layout or comprehensive.

Triangles are available in clear acrylic, tinted acrylic, acrylic with metal beveled edges, and solid metal. Because of their transparency, acrylic triangles let you see the underlying artwork. Their main disadvantage is that they can be damaged if you use them for cutting. Metal triangles are excellent for cutting but do not let you see the underlying artwork. A good compromise is a clear acrylic triangle with a metal cutting edge. There are also adjustable triangles that can be set at any desired angle.

Schaedler Precision Rules

Schaedler Precision Rules come as a set of two rules: one for measuring in inches and millimeters, one for measuring agate lines and picas and for sizing lines and solid dots (used for highlighting written or other information) called *bullets*. The first rule is calibrated from 1/62 to 12 inches on one side and from 0.5 millimeters to 30 centimeters on the other. The second is calibrated from 1 to 168 agate lines on one side and from 1 point to 72 picas and shows ½- to 30-point lines and 1- to 30-point bullets.

A special feature of Schaedler Precision Rules is that they are transparent, so you can place the rule right over whatever you are measuring and see through rather than having to place the rule alongside what you are measuring. Another feature is that the rules are made of a thin, flexible polyester film, making them suitable for measuring three-dimensional as well as flat objects (Fig. 3.12).

Type Gauge

A *type gauge* is a hard plastic rule with a scale for measuring agate lines, inches, picas, and 6-to-15-point type. It is used to character count, on manuscript copy, and to measure line spacing and column depths in layouts.

Circular Proportional Scales

A circular proportional scale is used to calculate enlargements and reductions of artwork, photographs, type, and layouts (Fig. 3.13). It consists of two different-sized plastic disks attached in the middle so that they rotate.

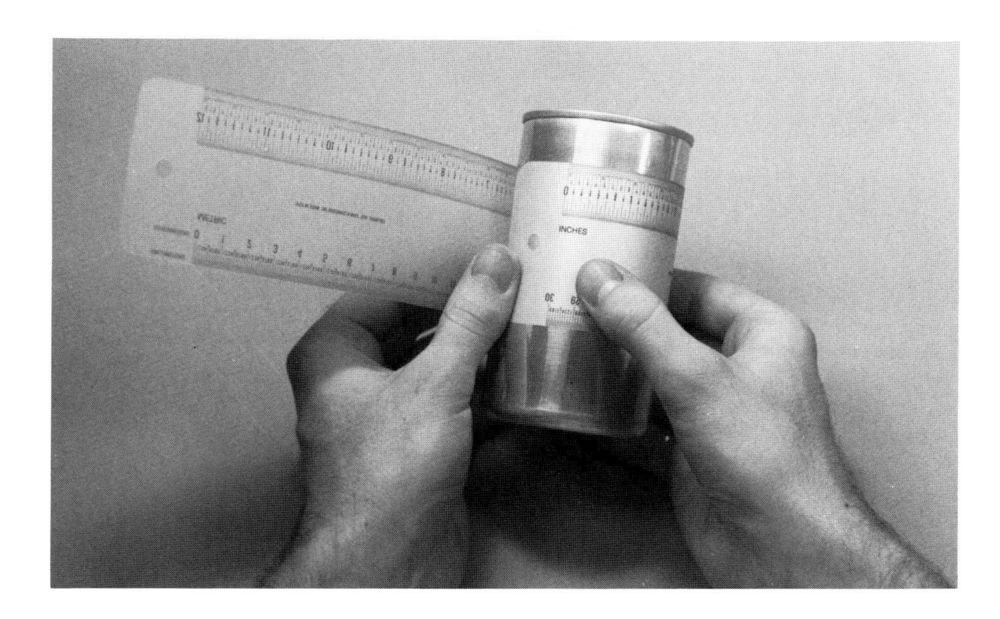

Figure 3.12 *(top).* Flexible Schaedler Precision Rules enable you to measure many three-dimensional objects.

Figure 3.13 *(below).* Circular proportional scale.

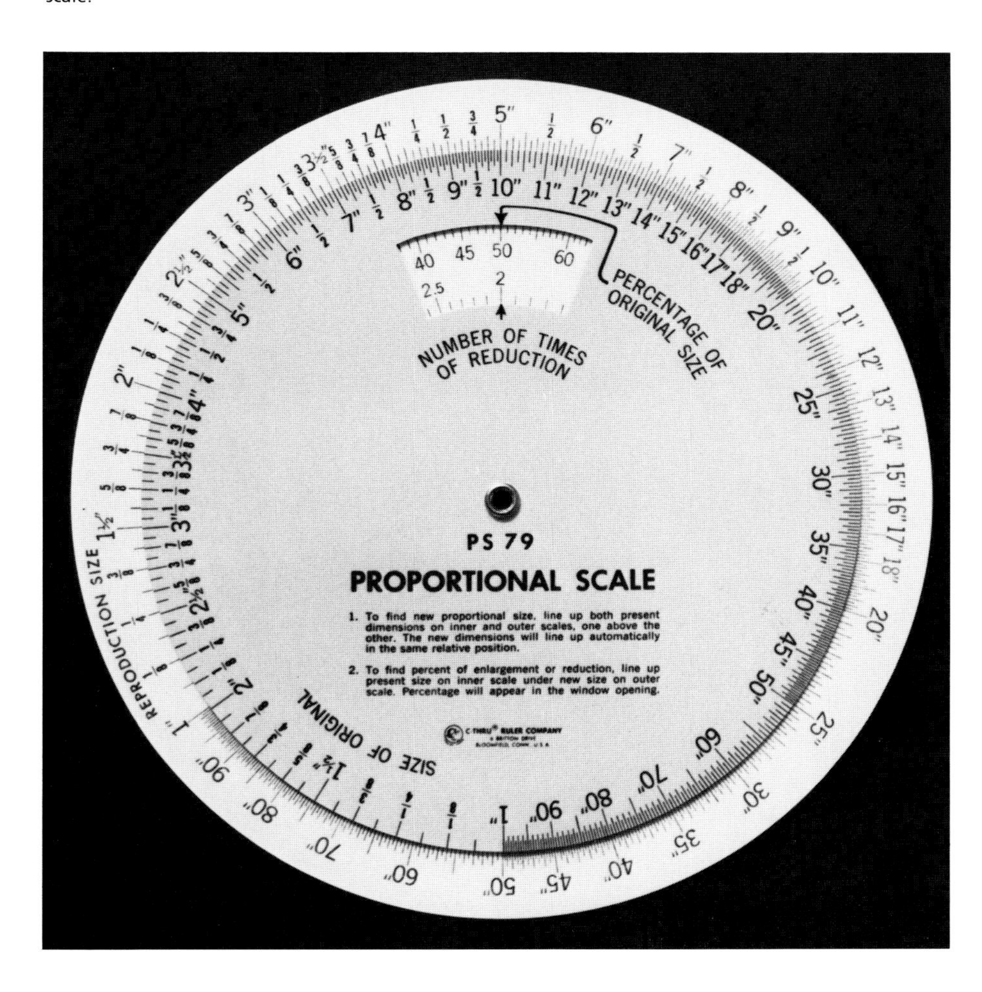

To use the scale, you locate the size of the original type or artwork on the smaller disk and the size you want it to be on the larger disk. You then rotate the smaller disk until the original size and the desired (or reproduction) size line up. A window cut out of the smaller disk reveals the percentage of the original size that will give you the desired size, as well as the number of times the original is being reduced.

Drawing Guides

As a graphic designer, you will need to draw shapes, letters, curves, symbols, and so on, neatly and accurately. To help you, drawing guides are available (Fig. 3.14). One type of drawing guide is the *template,* which is a piece of plastic that has shapes and symbols cut out so that you can trace them onto paper or other surface. Templates are available in hundreds of varieties, such as those for engineers or architects, or circular or hexagonal ones (Fig. 3.15). The templates you will most likely use are covered in the following discussions.

Another type of drawing guide is the *curve.* Two common curves—flexible curves and French curves—are described after templates.

Lettering Templates

Although limited in sizes and typefaces, lettering templates allow easy tracing onto layouts with pencil, fine-tip marker, or colored pencils. Typical typefaces include Block-Stencil, Old English Script, and Futuro. Numerals are provided with each typeface.

Ellipse Templates

Plastic ellipse templates are used in graphic design for rendering lettering artwork, or logos, for creating borders and decorative shapes and patterns, and, of course, for drawing uniform ellipses. Ellipses range from ⅛ to 4 inches and from 10 to 80 degrees. When you specify the

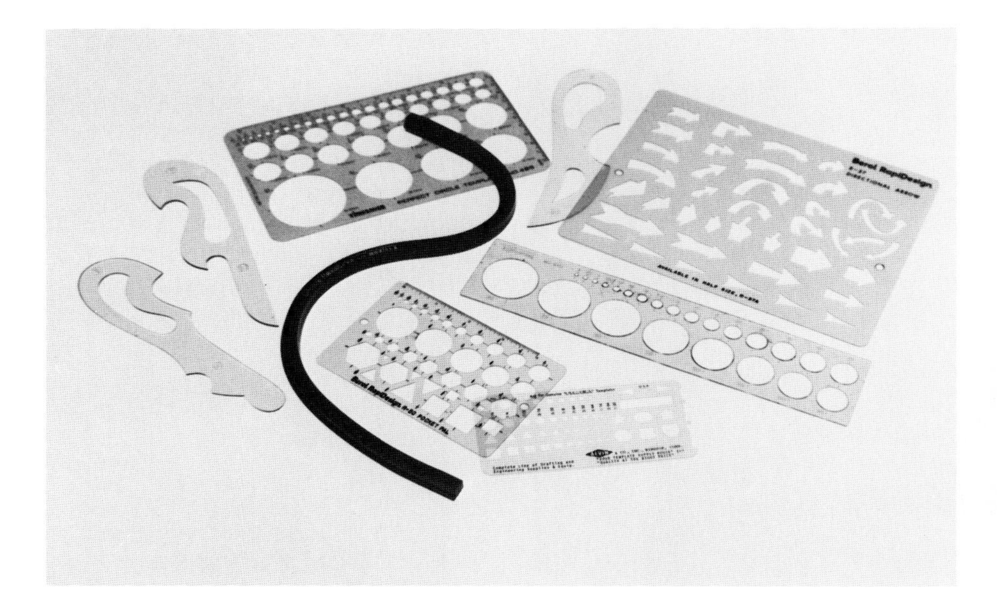

Figure 3.14 *(left).* Drawing guides *(left to right):* French curves, flexible curve, and assorted templates.

Figure 3.15 *(below).* Hundreds of templates are available to aid in the accurate drawing of common shapes, ellipses, lettering, human figures, and many other forms.

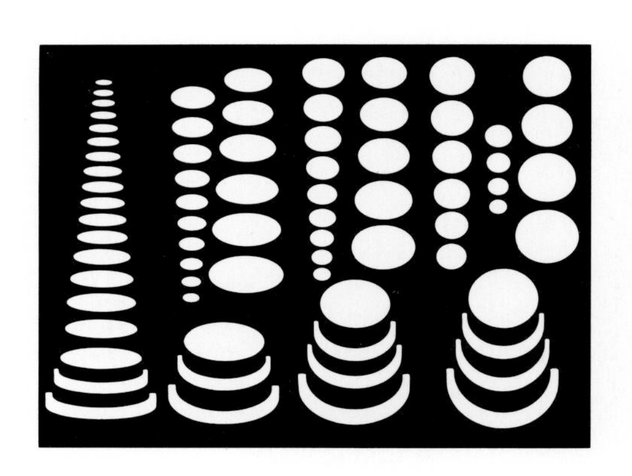

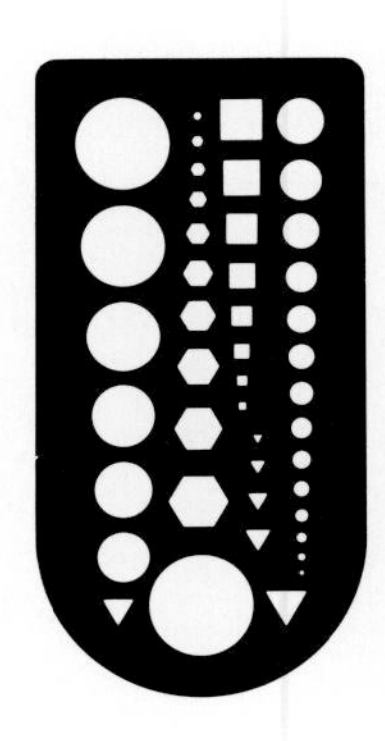

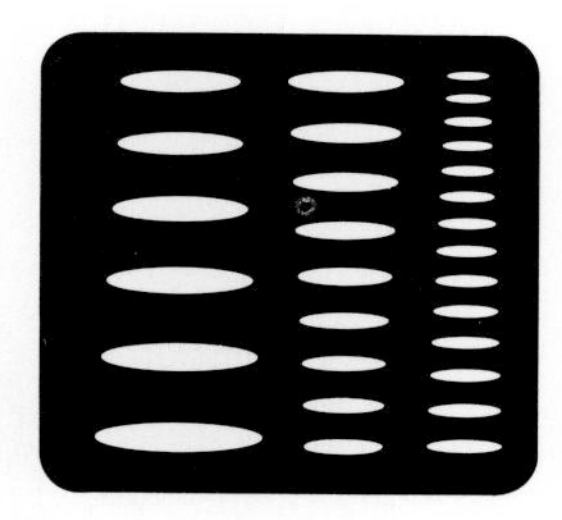

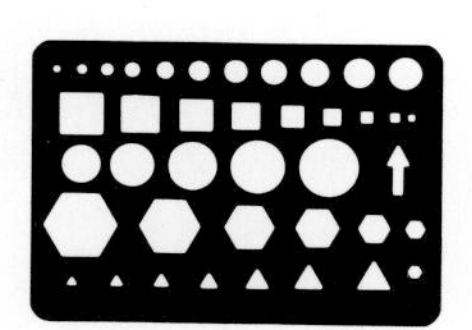

dimensions of an ellipse, you give its major axis in inches and its minor axis, which refers to its angle of foreshortening, in increments of 5 degrees. You can purchase ellipse templates individually, in sets of several, or as a complete ellipse template set.

Circle Templates

Available in increments from $\frac{1}{16}$ to $3\frac{1}{2}$ inches in diameter, circle templates are accurate guides for sketching or for finished inking in layouts and comprehensives. Although their sizes are limited, they are much faster to use than a compass and are easy to use with pencils, pens, and markers.

Flexible Curves

The *flexible curve* is an adjustable curved ruler that looks like a snake (see Fig. 3.14). Its unique design will let you bend it into any curve and use it on most surfaces without having it move around. Flexible curves are generally used in graphic design as a guide for drawing large, simple, irregular curves. They are excellent for finished hand-lettering and for drawing artwork and logos for comprehensives and advanced comprehensives. They are available in lengths of 12 to 30 inches.

French Curves

French curves are used to draw a wide range of irregular curves (Fig. 3.16). They are made of clear plastic with either a square or beveled inking edge. You can purchase these curves in sets of three or more. When using a French curve to draw similar shapes, make markings directly on the template to ensure consistent strokes (Fig. 3.17).

Drawing Instruments

In a field in which accuracy and precision can improve or ruin any design solution, having well-made, depend-

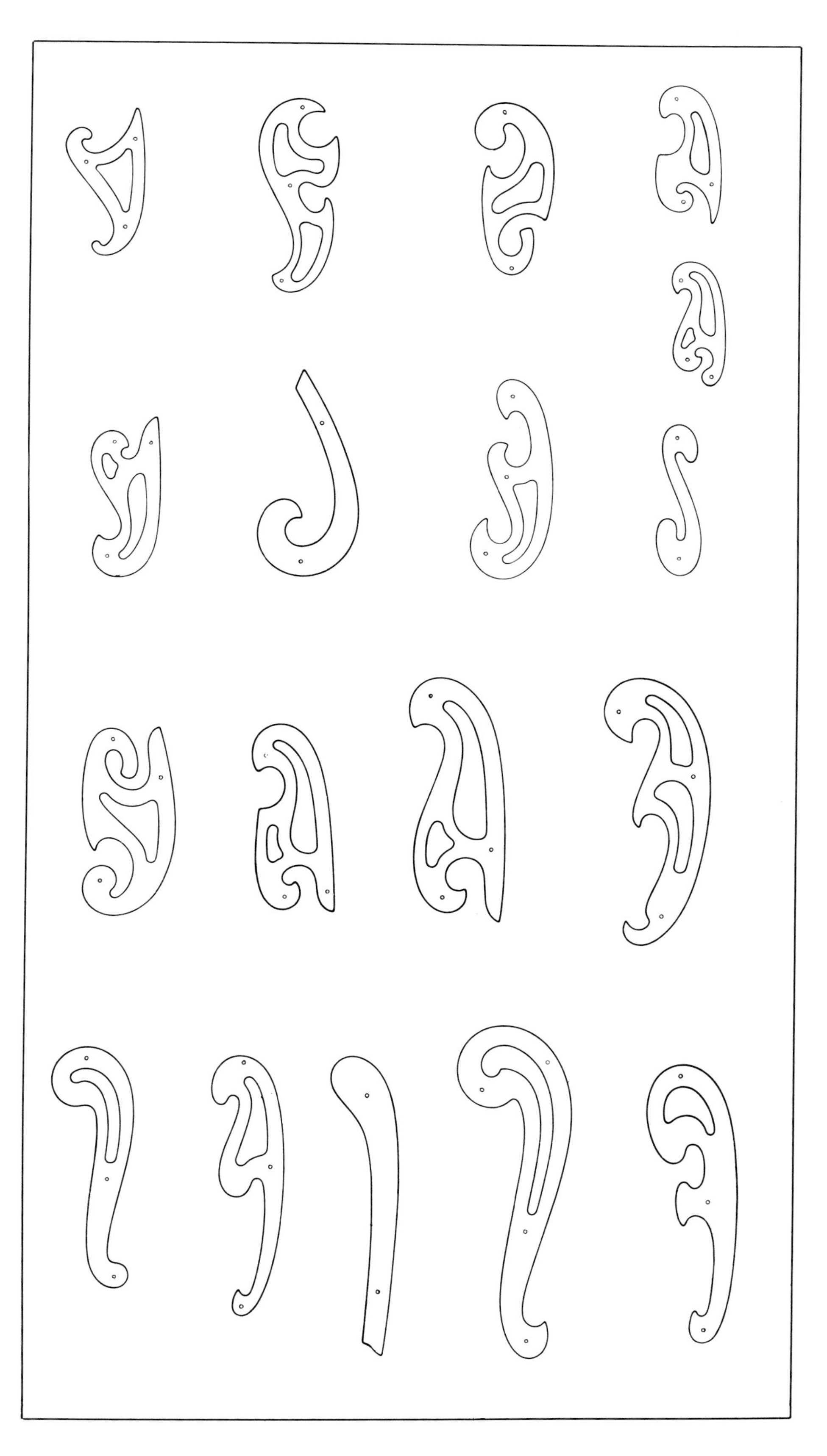

Figure 3.16. Assortment of French curves.

Figure 3.17. When drawing similar shapes or curves, make markings on the actual template to ensure that the strokes are consistent.

able tools on hand can make a big difference. Many designers are proud of their drawing instruments, as craftsmen are of their tools. Drawing instruments are used only in the comprehensive and advanced stages of a graphic design solution to achieve very finished results.

As with many drafting tools, there are two different grades—scholastic and professional. The budget you have will generally determine what grade you buy. Packaged individually and in sets, both grades will help you produce clean, presentation-quality comprehensives. Proper maintenance of any drawing instrument will ensure that it will last for a very long time.

The sets contain tools, such as dividers and ruling pens, as well as compasses with interchangeable leads, and needles and blades. The lower-priced sets are usually adequate for the beginning student to become familiar with each piece. Additional pieces can be added later on.

Compasses

A *compass* is used to draw, ink, or cut perfect circles. All compasses have two "legs," one with a sharp point to grip the surface and the other with a lead holder, ruling or technical-pen tip, or cutting blade.

Compasses are available individually or in sets containing a wide range of sizes and styles (Fig. 3.18). The choice of compass is determined by the size of circle you want to draw. Use a drop compass to make small circles up to a half inch in diameter; these have both graphite and pen points. A bow compass, which is the most common, comes in two sizes: small, to draw circles up to 3 inches in diameter, and large, to draw circles up to 12 inches in diameter. An extension bar can be used with the large bow compass to draw very large circles (usually larger than 12 inches). When you draw large circles with a bow compass, make sure both the leg and the drawing tip are perpendicular to the work surface. To draw extremely large circles, use a beam compass.

Beam Compass Cutters

You can use the combination beam compass cutter to draw or cut large circles (Fig. 3.19). Many models can be fitted with lengthening bars (for bigger circles), leads, pens, brushes, or knife blades. Unlike smaller compasses, beam compasses require two hands to operate. One hand holds the needle point and the other rotates the arm. Small blades can also be inserted into a bow, which can then be used to cut circles.

Dividers

Dividers look like compasses except that each of their legs is needle-shaped. A divider is used for transferring or duplicating measurements from one surface to another quickly and accurately without having to measure each time. To transfer a measurement, first set the divider's legs to match the distance on the original, then move the divider to the new surface and prick the surface with the sharp points of the divider. The distance between the two marks will be identical to that on the original, and more accurate a measurement than that taken from a ruler.

Ruling Pens

Ruling pens are used for drawing clean, accurate lines of a consistent thickness. The pens have a handle and a blade opening, with an adjustable screw on the side that opens and closes the opening. Ink or any other liquid medium is placed in this opening with an eyedropper or brush. Once filled, the blade opening can be adjusted to draw lines of varying thicknesses. When using a ruling pen, always use it with a straightedge to make sure you get straight lines of uniform width. When ruling, place the pen along this top edge and drag it slowly, at even speed, to the right edge, maintaining even pressure.

Media used with ruling pens include black India ink, other inks, liquid watercolors, and designer's

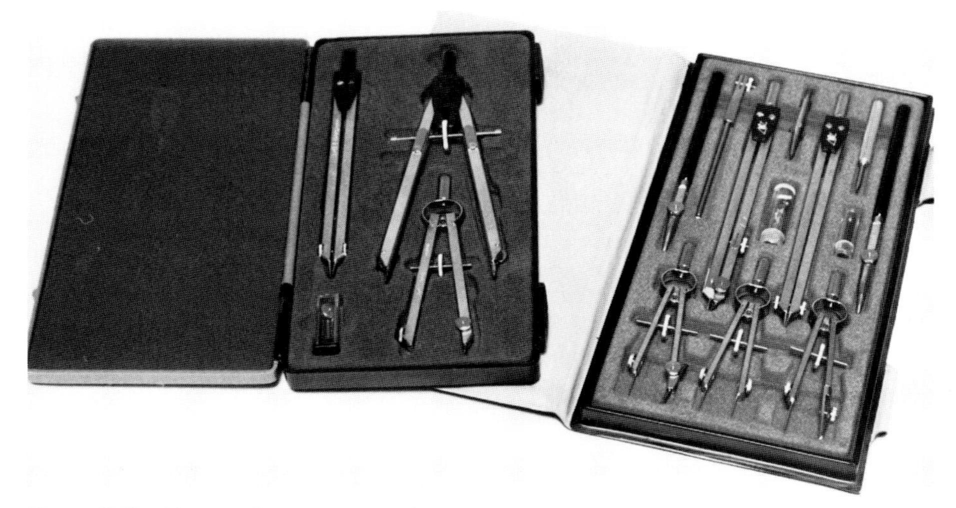

Figure 3.18. Two popular compass sets: student and professional.

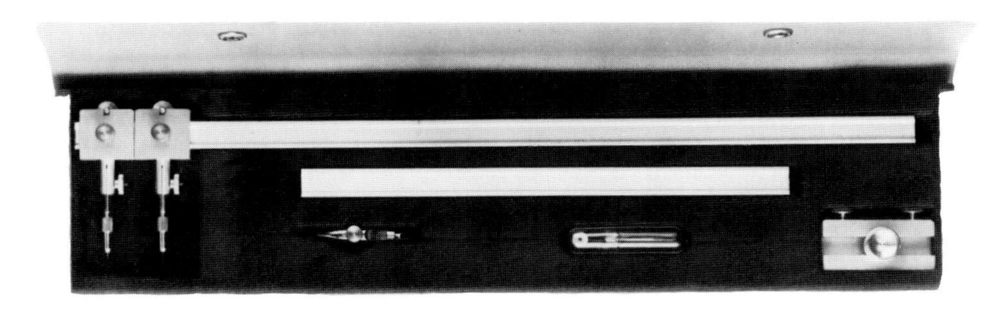

Figure 3.19 *(above).* Beam compass cutter set.

Figure 3.20 *(below).* Cutting tools *(left to right):* Lightweight stainless-steel scissors, single-edge razor blades, art knife (with #11 blade), and utility knife.

colors (thinned with water). To fill ruling pens, use the built-in ink dropper that comes with India ink, wipe a brush saturated with color against the blade, or use a specially made black India ink cartridge.

When buying a ruling pen, make sure the blades are sharp, strong, and noncorrosive so they will not become worn with extended use. A calibrated adjusting screw can also help you maintain even rule thickness.

Cutting Tools

Seldom does a day go by when a designer does not use a cutting tool to perform a task in the layout or comprehensive stage of a design assignment. Although different cutting tools can be used for the same task, you should get into the habit of using the cutting tool most aptly suited to the task at hand to save time and avoid accidents. For example, although an art knife is perfectly capable of cutting through mat board, it is faster and much safer to use a sturdy utility knife to do the job. Figure 3.20 shows a variety of cutting tools.

Art Knives

A designer's most frequently used cutting tool for light-to-medium work is the art knife. A popular model, the X-Acto, has a handle with a screw-type metal locking device that holds the blade when you want to cut and instantly releases it when you want to change it. It is used with surgically sharp steel blades, which are available in many sizes and shapes.

The blades used most often by graphic designers are the #11 blade for general cutting work (Fig. 3.21) and the #16 for cutting film and curves. The blades can be sharpened to their original cutting edge with a sharpening stone lubricated with a light oil.

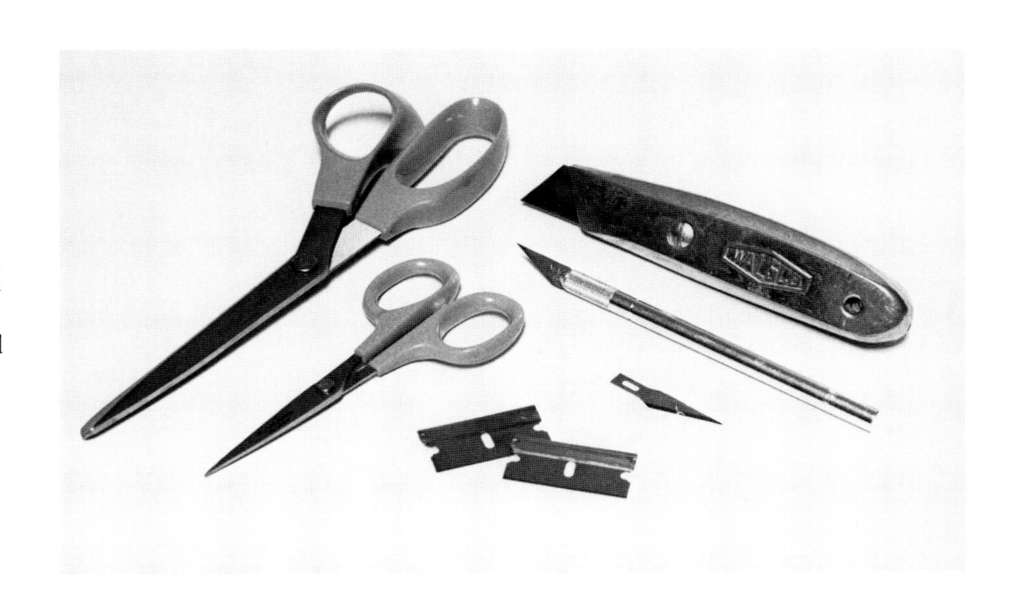

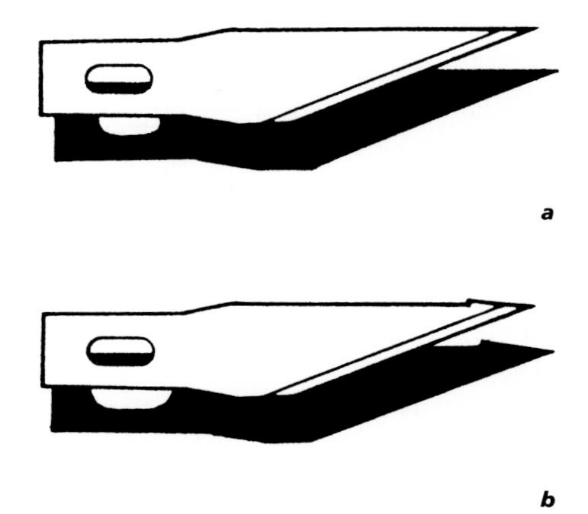

Figure 3.21. *a.* Standard #11 blade. *b.* #11 Modified.

Another type of art knife is one with a swivel-type handle that houses tiny low-angled blades for precise cutting of friskets (stencils or masks that are cut out of paper or film and used to block out or protect artwork) and freehand curves without lifting your hand or adjusting your fingers.

Utility Knives

The utility knife (Fig. 3.20) is a heavy-duty large-handled knife that enables designers to apply enough pressure to cut through thick materials such as mats and art boards. Refill blades are usually stored inside the handle for safety and convenience. The most popular version is the locked-handle version, which keeps the blade from accidentally retracting during cutting. Other types have three-position retractable blades for cutting.

Single-edge Razor Blades

Designers use industrial grade single-edge razor blades to score or cut heavy cardboard and to shave away wood on pencils before sharpening them.

Sharpening pencil points with a single-edge razor blade lets you achieve a wide variety of pencil-point shapes. If your layout requires fat, uniform-width lines to simulate headline or text type (body copy), sharpen a drawing or sketching pencil to a chisel point after removing the wood around the lead. If a needle-sharp tip is required, such as for indicating text type with loops, rub the pencil tip over a sandpaper pad while at the same time twisting the opposite end of the pencil. Before sharpening with a single-edge razor blade, it is important to remove wood, exposing as much lead as possible without weakening it, so as not to have to constantly resharpen your pencils. Razor blades are also used to cut chisel points for text and display type.

Scissors

Designers use sharp, general-purpose scissors for cutting paper, fabrics, and any unmounted material. Although some scissors will cut through many thick materials, such as mounting boards, this takes longer and will dull the scissors. Fiskars, a popular all-purpose scissor, is available in many lengths and is made of lightweight stainless steel with comfortably molded finger grips that minimize fatigue and discomfort.

Brushes and Color Applicators

Whether brushes and color applicators are used to render small type or large background areas, they are very important tools in making hand-rendered comprehensives (Fig. 3.22).

Good brushes can last a long time if you take care of them. To preserve your brushes, use a specific brush for each medium since no matter how well you clean a brush, something can become trapped in the ferrule (the ring of metal on the brush). For example, have a brush that you use only with black India ink and one that you use only with white ink for retouching. Another way to take care of brushes is never to allow any

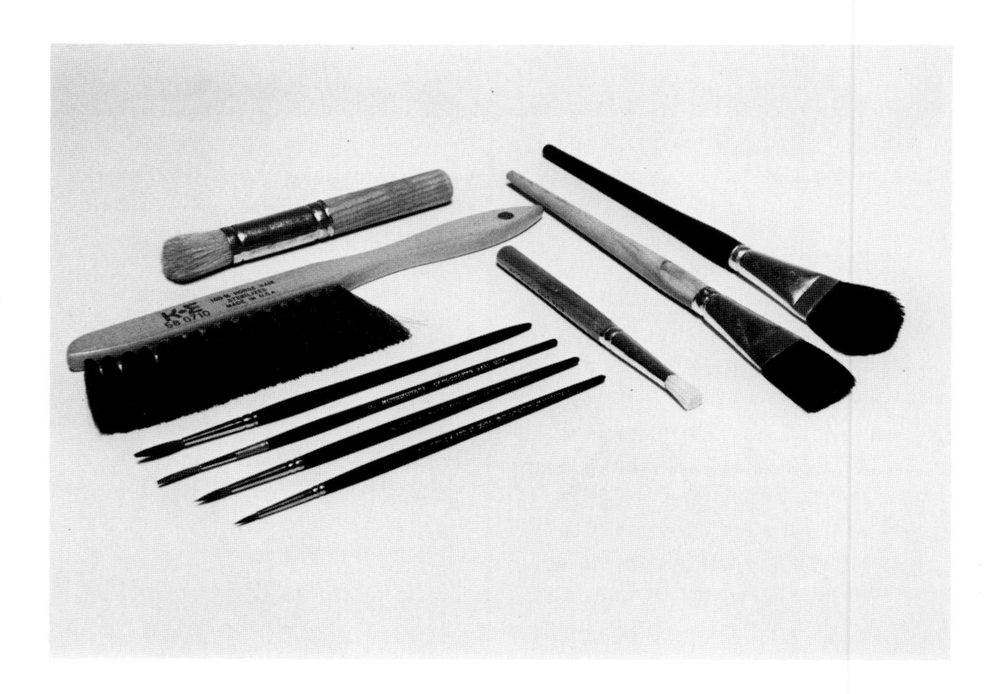

Figure 3.22. Brushes *(counterclockwise from top):* Stencil brush (large), dusting brush, lettering brushes (#10 and #5), watercolor brushes (#3 and #1), stencil brush (small), red sable one-stroke brush, and flat oval-shaped watercolor brush.

medium to harden or dry in the brush since that will cause the hairs to become brittle and break off.

When you finish using a brush, wash it immediately in warm water with mild soap until the medium is removed. Then with saliva, reshape the hairs to their original point.

Watercolor Brushes

The most widely used brush in the graphic design studio is the pointed watercolor brush. The finest-quality pointed watercolor brushes are made of red sable. These brushes hold a large amount of color, maintain a sharp point, and return to their original shape after each stroke.

A useful feature of red sable brushes is that a wide variety of strokes can be made with them, depending on the way the brush is held. If you create a tip, hold the brush perpendicular to the work surface, and use a small amount of color, you can produce fine lines for lettering or retouching. If you hold the brush at a slant and use a large amount of color, you can produce wide lines or broad strokes, called washes, for creating a background or filling in large areas.

The most popular red sable brushes are the Winsor & Newton Series 7 and Grumbacher No. GR 197. They are available in sizes from 000, which has a needle-point tip, to 14, which has a wide tip. The sizes most often used for graphic design work are 00, 1, 3, and 6. These can be used for most tasks, ranging from cleaning up final lettering and artwork to filling in large areas and painting the edges of elements mounted on a mechanical. If you need larger brushes for background washes or large areas, rather than purchase an expensive No. 12 or No. 14 red sable pointed brush, you can use a red sable one-stroke brush (also used for sign-writing, lettering, etc.) or flat oval-shaped watercolor brush.

Lettering Brushes

Lettering brushes have round ferrules and clean, square tips, which make them especially useful for one-stroke comp lettering on comprehensives. They are available in many sizes in red sable, ox hair, and sabeline, and come with handles of varying lengths.

Airbrushes

An airbrush is an excellent graphic-design tool for making illustrations, lettering, and backgrounds, and for retouching. You can use it with a limitless range of colors and media and can create fine grain and an assortment of line widths. Basically, an airbrush uses air pressure to transfer a finely atomized spray of a liquid medium from a reservoir onto artwork. It contains two valves: One lets the liquid medium be sprayed through the nozzle; the other controls the amount of air that enters the airbrush, which in turn controls the amount of medium that comes out of the reservoir.

There are many models of airbrushes that offer a variety of capabilities; the most commonly used in graphic design are the Paasche Model V and the Thayer and Chandler Model A. These models are double-action airbrushes, which means that you use your fingers to move a lever either forward or backward thus controlling the flow of media and air. This feature enables you to operate the airbrush with one hand and gives you control over what comes out.

All airbrushes need a source of compressed air to operate. These include portable air compressors (Fig. 3.23), portable pressure-tank units, and carbonic gas cylinders. Some portable air compressors are made specifically for use with an airbrush, but you can use any air compressor that can be regulated to between 25 and 35 pounds and has a filter to remove dirt, water, and oil from the air. Portable pressure-tank units such as the Paasche No. 2 Pressure Tank and the Badger Propel will supply enough air pressure for one to three hours of operation, depending on the unit's size and model.

You can use an airbrush with many different media, including liquid dyes, watercolors, drawing and India inks; *designer's colors* or *gouache* (thinned with water), and photo-retouching and oil paints. When using designer's colors and retouch-

Figure 3.23. Paasche Model VI airbrush with compressor.

ing colors, use a thinner consistency than for regular brush application. When using waterproof ink, thin it out with household ammonia, which can also be used to clean the airbrush. When you are using dyes, clean the airbrush with laundry bleach. After using these harsh cleaning liquids, be sure to flush out the airbrush with water so that it will not become damaged. There is also a commercially prepared concentrated airbrush cleaner, Com-Art, which when diluted with water will clean dyes, inks, acrylics, gouache, watercolors, and Com-Art retouching colors. It is noncorrosive, rinses off with water, and can be used to clean regular brushes as well. Its unique features make it an excellent layout and comprehensive-rendering tool.

LetraJet

LetraJet is a tool from the Letraset company that you can use to achieve airbrush effects (Fig. 3.24). Although it is not as good as an airbrush in some ways, it does have unique features that make it excellent for making layouts, comprehensives, and advanced comprehensives.

The LetraJet is used with fine-nib Pantone® markers (you cannot use broad-nib markers with it), and a useful feature is that you can change colors almost instantly just by changing the marker. There are 108 fine-nib Pantone by Letraset markers, which are coordinated to the Pantone Matching System to match Pantone publications, papers, films, and printing inks used in the graphic arts. The Pantone Matching System was developed so that designers and printers could have the same tools to help keep colors consistent when rendering a layout from thumbnails to advanced comprehensives to printing. Another advantage of the LetraJet is that it is easier to learn to use than an airbrush. You can master techniques for creating highlights, silhouettes, tonal renderings, dropouts, and color-on-color effects in a shorter time than you can with an airbrush.

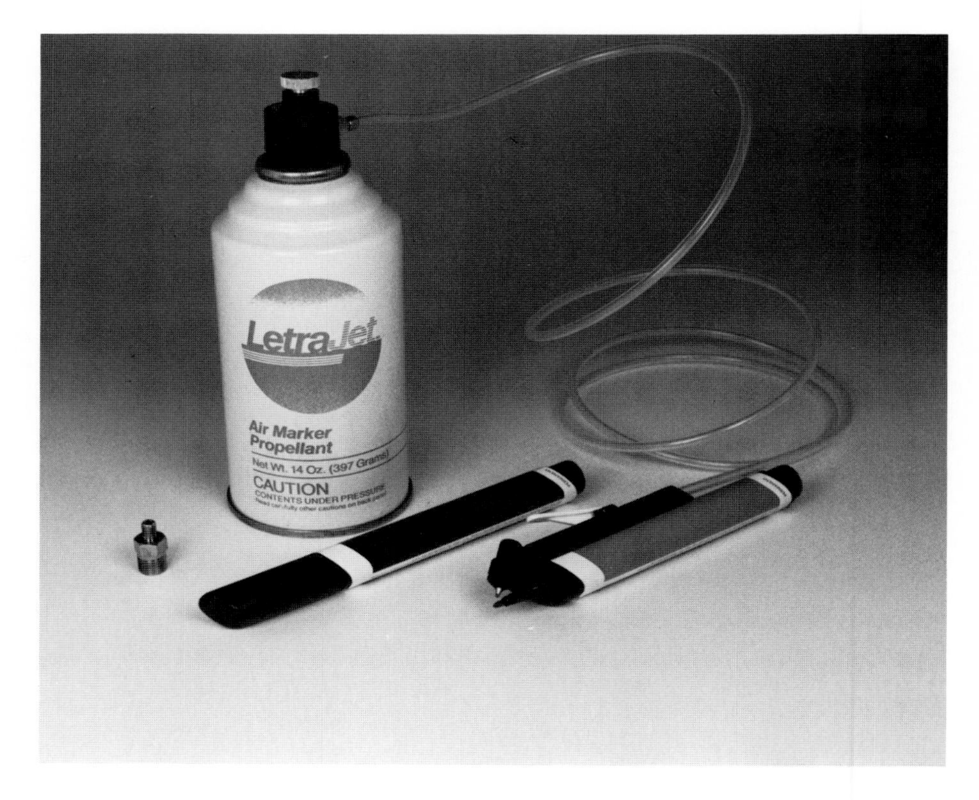

Figure 3.24. LetraJet, a portable alternative to the airbrush for use with comprehensives.

The LetraJet needs a power source, such as the LetraJet Air Marker Propellant, other aerosol propellants such as those used for airbrushes, or compressors that provide a maximum pressure of 60 pounds per square inch. The LetraJet comes with an adapter that easily connects it to a compressor.

Dusting Brushes

Long-handled general-purpose dusting or drafting brushes made of natural hair are used for removing particles and eraser crumbs from most work surfaces. They are the most efficient tool you can use to clean up during or after a project since they usually will not smudge your artwork.

Pre-Val Sprayer

The Pre-Val Sprayer is a unit that will finely atomize and spray most liquid media such as acrylics and designer's colors that have been diluted with water (Fig. 3.25). It can also be used to spray shellac, varnish, and enamel. Since the sprayer con-

Figure 3.25. Pre-Val Sprayer.

tains an airtight clear bottle with a screw-on cap, you can also use it to store colors for any length of time. This feature is especially useful for colors you have mixed yourself and may want to refer back to later.

Pencil- and Lead-Sharpening Tools

Pencils and leads can be sharpened using a variety of methods: The one you choose depends on the desired effect or the intended use (such as whether your layout needs fat, uniform-width lines to simulate text type). Also the shape and content of the pencil lead will determine the best method for sharpening. For example, rectangular-shaped lead housed in an oval wood holder will need to be sharpened by hand using a single-edge razor blade and a sanding pad; lead rods held in lead holders have to be sharpened with a lead pointer if you wish to get a needle-nosed point. Figure 3.26 shows the most widely used tools for sharpening pencils and leads.

Portable Pencil Sharpeners

A hand-held portable pencil sharpener enables you to sharpen pencils anywhere, in a classroom or library or on a bus. Some of these pencil sharpeners are made with a built-in storage well for collecting pencil shavings; others are capable of making points on leads.

Electric and Battery-operated Pencil Sharpeners

The advantages of electric and battery-operated pencil sharpeners are that they save time, produce uniformly tapered pencil points, can be operated with one hand, and are effective for media such as Prismacolor colored pencils, which because of their high wax content tend to wear down rapidly and need constant resharpening to maintain an adequate point.

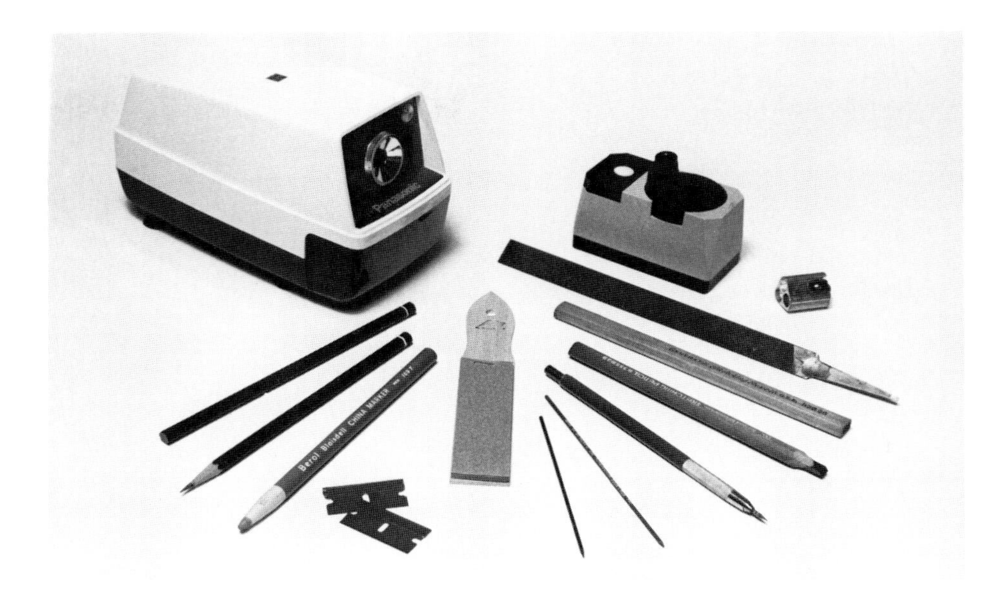

Figure 3.26. Pencils and sharpeners *(counter clockwise from left to right):* electric pencil sharpener, 2H drawing pencil, colored pencil, grease pencil, single-edge razor blades, sanding pad, graphite lead rod, blue lead rod, lead holder, flat sketching pencils, flat bastard file, portable sharpener, and lead pointer.

Lead Pointers

A lead pointer is the only tool available for sharpening drawing leads that are housed in mechanical lead holders. The lead pointer grinds colored or black graphite leads against a multiple-edged cutting wheel to produce long, needle points or short, stubby points. The pointers are available in hand-held portable models and desktop models that can be mounted on a drafting table for single-handed operation.

The operation of a lead pointer is simple. First, expose a portion of lead so it extends about one-half inch beyond the claws of a mechanical lead holder and lock it into place. Next, insert the mechanical lead holder into the pointer (lead first), grip the lead holder firmly, and spin it slowly (maintaining a downward pressure) until the resistance gives way and the point is sharp.

Sandpaper Blocks

A sandpaper block consists of several sheets of fine-grade sandpaper attached to a wooden block. It is used for creating points and shaping pencil leads and for keeping beveled and chisel points clean and sharp. To create a point on a pencil lead, hold the pencil lightly against the sandpaper pad at an angle that will produce the best point for its intended use. If you want a long, tapered point, hold the pencil at the long angle almost parallel to the block. If you want a short, wider point, hold the pencil at a greater angle to the block. Rotate the pencil slightly while moving the point back and forth over the sandpaper surface.

Flat Bastard File

A flat bastard file is used to create chisel points on flat sketching pencils (*see* Chapter 4, Making a Chisel Point). Available in any hardware store, this file, which has a coarse cross cut, is flat on both sides and will not clog as easily as sandpaper. Also, it can be wiped clean with a rag and reused. The flat bastard file, which is constructed from flat sturdy metal, is designed so that it can be held away from the body with one hand while the other presses the pencil tip onto the file at a 45-degree angle and moves it up and down, parallel to the direction of the file.

Tape Dispensers

Convenient tape dispensers come in many sizes, shapes, and materials. It is important to pair up the type of dispenser with the correct type of tape and vice versa. Avoid dispensers that will only hold tape rolls with one-inch (in diameter) cores since white artist's tape typically comes in three-inch cores. A one-half to one inch-width tape will serve most needs.

Desktop Dispensers

Desktop dispensers are generally made of plastic and metal and are weighted so they will not move when tape is being dispensed with one hand. They are available in one and two capacity sizes; the two-capacity dispenser is best because graphic designers use white and transparent tape most often.

Clamp-on Dispensers

Clamp-on dispensers are attached to the top or side of the desk. They allow single-handed dispensing and keep the top of the desk free for working.

Burnishers

Burnishers are used in the graphic design studio mainly to press down adhesive-backed or pressure-sensitive lettering, paper artwork, and colored and patterned film overlays to advanced comprehensives. Figure 3.27 shows several types of burnishers. They let you work quickly, are easy to use, and reduce the amount of damage to the background surface and the items you are transferring. They also help materials that have already been adhered to stay adhered. It is a good idea to have a wide selection of burnishers to handle all your needs in the preparation of comprehensives. *Note:* Colored film should always be burnished after it has been adhered in order to bring out its actual color.

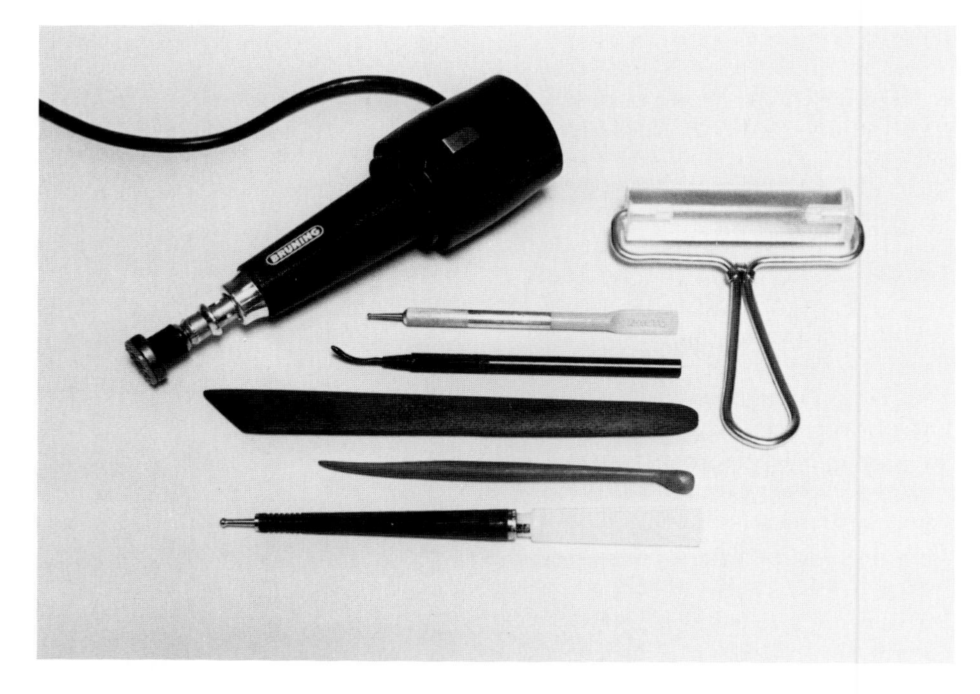

Figure 3.27. Burnishers *(top to bottom):* electric burnisher, Lucite roller, X-Acto ball burnisher, Letraset spoon-tip burnisher, wooden burnishers (large and small), and Chartpak adjustable burnisher.

To *burnish* means to press down paste-up materials (such as type) colored papers, photographs, or adhesive-backed colored or patterned film and transfer type such as Letraset. Items are usually burnished only in the advanced comprehensive stage because this stage usually requires materials that cannot be duplicated with the hand-rendering techniques reserved for comprehensives.

Manual Burnishers

Manual burnishers come in a variety of shapes and materials, each of which is used for a specific purpose. For example, a fine-tipped wood or metal burnisher is suitable for pressure-sensitive lettering and other delicate work; a wide chisel-tipped plastic burnisher is excellent for transferring large letters and film overlays quickly. Popular brands of burnishers are Chartpak, Letraset, and Zipatone.

Power Burnishers

A power burnisher is a rotating head that can be attached to most standard electrical or battery-operated electric erasers (see Fig. 3.27). With it you can apply rubdown transfers and I.N.T.'s from their camera sheets onto advanced comprehensives quickly with gentle, uniform pressure.

Lucite Rollers

Lucite rollers are used to flatten artwork or paste-up materials or to burnish film overlays, adhesive-backed paper, artwork, and photographs onto surfaces such as comprehensives permanently (see Fig. 3.27). With only slight pressure applied, the roller's smooth clear, Lucite surface lets you eliminate lumps, creases, and ripples and will not stick to wax, rubber cement, or your artwork.

Color Specifiers

When ordering colored papers, films, rub-down transfers, or printing inks, several methods can be used to specify color. One is to attach any color swatch to the artwork or mechanical and have it matched by relying on a printer's color-mixing ability. A more accurate method is to ask to mix a color from an established color matching system, such as Pantone. The Pantone Matching System is the one most often used. This standardized color system, available worldwide, will ensure accurate color reproduction throughout a design project—from layouts to advanced comprehensives and printing. In addition to the Pantone Color Specifier book (which contains color chips of printing inks), there are a number of related products such as color markers, adhesive-backed overlay films, and colored papers that allow the designer to accurately control color reproduction throughout a job.

Figure 3.28. Pantone color-specifying materials *(clockwise from top):* Pantone Color Specifier and inside pages, Pantone by Letraset Color Overlay Selector, and Pantone Coated Paper Selector.

Pantone Color Specifier

The Pantone Specifier is a dictionary of 747 colors that provides accurate reference for any printing ink color in the Pantone Matching System (Fig. 3.28). Each color is shown on both coated and uncoated paper stocks in a loose-leaf binder. The pages have been perforated into chips for easy removal. You can attach these chips to your artwork, layouts, rub-down transfer boards, or mechanicals for true color comparison and matching. Individual pages and sections can be replaced when they are used up.

Other Color Specifiers

Other color specifiers, many of which are also part of the Pantone Matching System, are available. They include coated and uncoated colored paper selectors, matte film selectors, and colored overlay film selectors. Another color matching system that has gained popularity in recent years is the Toyo Color System from Japan.

Note Pads

Another overlooked tool is the note pad. You should have an ample supply of note pads in the studio to use during all phases of a project. Use them to record instructions at a design briefing, as a daily journal to jot down ideas that come to mind in the course of a project, and so on.

Post-it Note Pads

Post-it note pads are removable self-stick notes available in 1-by-2-, 2-by-3- and 3-by-5-inch pads of 100 sheets (Fig. 3.29). Graphic designers use them to write instructions and corrections on photographs, type galleys, manuscripts, and comprehensives without damaging the surfaces. They can also be used for markers and flags, in magazines and books.

When you are rendering backgrounds or broad areas on layouts with markers, you can attach the larger Post-it sheets along the outside to let you make long, uninterrupted strokes without ink build-up. Immediately after you finish the rendering, remove the sheets, and you will have straight, clean edges and a uniformly covered background.

Figure 3.29. "Scotch" Brand Post-it Note Pads perform numerous functions in the graphic design studio.

MATERIALS AND MEDIA

The true magic of developing an idea from thumbnail to advanced comprehensive begins with the designer's imagination and becomes visible when ideas take form on paper. The materials and media, and of course the skill required to develop an idea throughout the design process, help the designer move the project through its course.

Pencils

Pencils have the largest role of any tool in the design studio. In the earliest layout stages of a design assignment, the pencil is usually the first item a designer will pick up when starting. Pencils are a very flexible medium because in most cases the marks can be removed by erasing. This feature allows for a great deal of flexibility in making changes and revisions, encouraging experimentation in the thumbnail or layout stage.

Soft-lead graphite and soft-wax-type colored pencils are used in these early stages because they are capable of rendering quick, broad strokes. Mechanical lead holders and pencils containing hard leads are generally used in the more finished comprehensive stages where accuracy is important.

Drawing Pencils

Most drawing pencils are wood with graphite (a variety of carbon) centers. Drawing pencils are available in seventeen degrees of hardness, ranging from 6B, which is very soft and produces thick black lines, to 9H, which is very hard and produces faint lines (Fig. 3.30). A few soft, medium, and hard drawing pencils will provide enough variety for most graphic-design work.

Drawing Leads and Lead Holders

Drawing-pencil leads used in combination with lead holders are generally used for producing thin, faint lines in

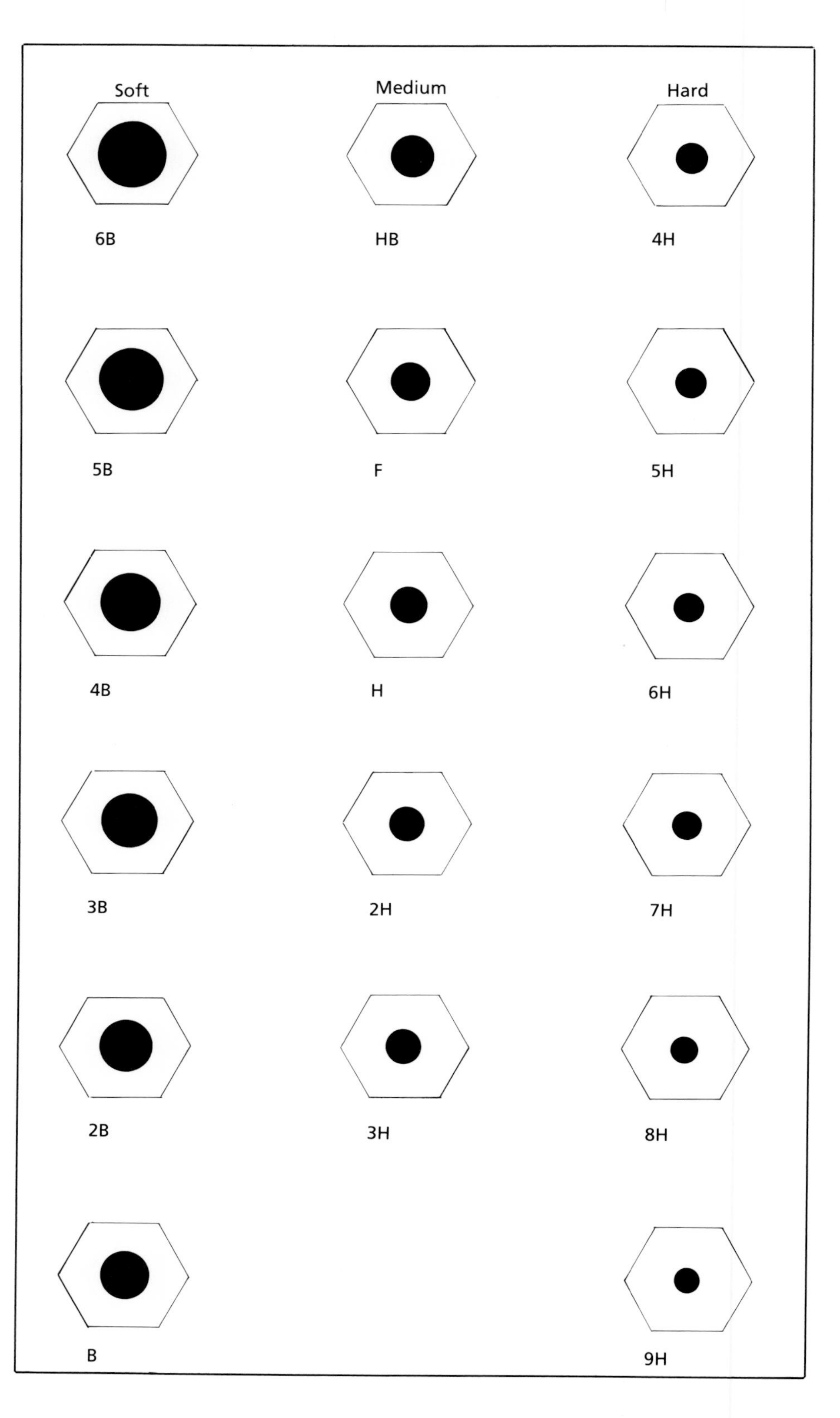

Figure 3.30. There is a wide variety in the diameters of drawing pencil leads. Note that the hardest lead has the smallest diameter (magnification: approx. × 3).

finished layouts, comprehensives, advanced comprehensives, and for indicating accurate guidelines. Graphite drawing leads are available in many degrees of hardness, ranging from the very soft 6B used to produce heavy black lines to the very hard 9H used most often for guidelines on mechanical artwork.

Lead holders are made of metal and plastic or a combination of both. The push-button action on the end of the barrel controls the jaw clutch grip that holds the lead in place when working and allows the lead to be pulled out for sharpening or pushed in for transporting or protection.

The main advantage to using drawing leads and lead holders over conventional pencils is the long, sharp point, which can be made by sharpening the lead with a lead pointer when accuracy is essential to a layout, a comprehensive, or mechanical artwork. Otherwise, any other standard pencil will work as well for most uses.

Flat Sketching Pencils

Flat sketching pencils, which are larger than drawing pencils, have rectangular graphite centers used for covering large areas quickly and for lettering. They are available in three degrees, 2B, 4B, 6B, and in two brands, General's and Koh-I-Noor sketching pencils. The General's pencil is rectangular and has a large graphite center. The Koh-I-Noor pencil is oval-shaped and has a slightly smaller graphite center. Because the lead is square (rather than rounded, as in most pencils), it allows you to make a wide, even stroke when lettering in a layout.

Colored Pencils

There are many types of colored pencils used in the graphic-design process. Prismacolor, which has a soft, wax-based center, is made from brilliant pigments and is used primarily for filling in large areas or for thick, intense coverage. These pencils can be sharpened to a fine point but tend to wear down rapidly because of their high wax content. Prismacolor colored pencils are available in sixty colors and are waterproof so you can use markers over them.

Verithin colored pencils have a thin, hard noncrumbling center in a hexagonal wooden shaft. They are used for fine line work and tend to retain their points longer than Prismacolor pencils.

Other types of colored pencils are erasable. Some can be used dry or blended with water, and others can be blended with a blending stump or your finger. A blending stump is a round, thin paper, wrapped and tightly rolled, which has a point on one end used for blending pastels, charcoal, crayons, and pencils.

Nonreproducing Pencils

Blue nonreproduction (or "nonrepro" blue) pencils and rods that create lines that will not reproduce when photostatted are used to mark up guidelines for mechanicals and artwork.

China-marking Pencils

China-marking pencils, also called grease pencils, are soft-wax, brightly colored pencils that are used to write on glossy surfaces such as plastic, glass, and metal. They are mostly used for indicating crop marks (showing areas to be included or taken away) and other special instructions on photographs. Pencil marks can be removed with a facial tissue moistened with rubber-cement thinner.

Erasers

For almost every ink pen or pencil there exists some form of eraser to eradicate almost any accident or mistake made. Having a selection of ink and pencil erasers will make quick clean-ups and deletions that much faster.

Pencil Erasers

Soft rubber pencil erasers are made specifically for erasing pencil lines without smudging (Fig. 3.31). Their disadvantages are that they are too abrasive for some papers and leave crumbs on the working surface. Recommended brands are Pink Pearl and RubKleen.

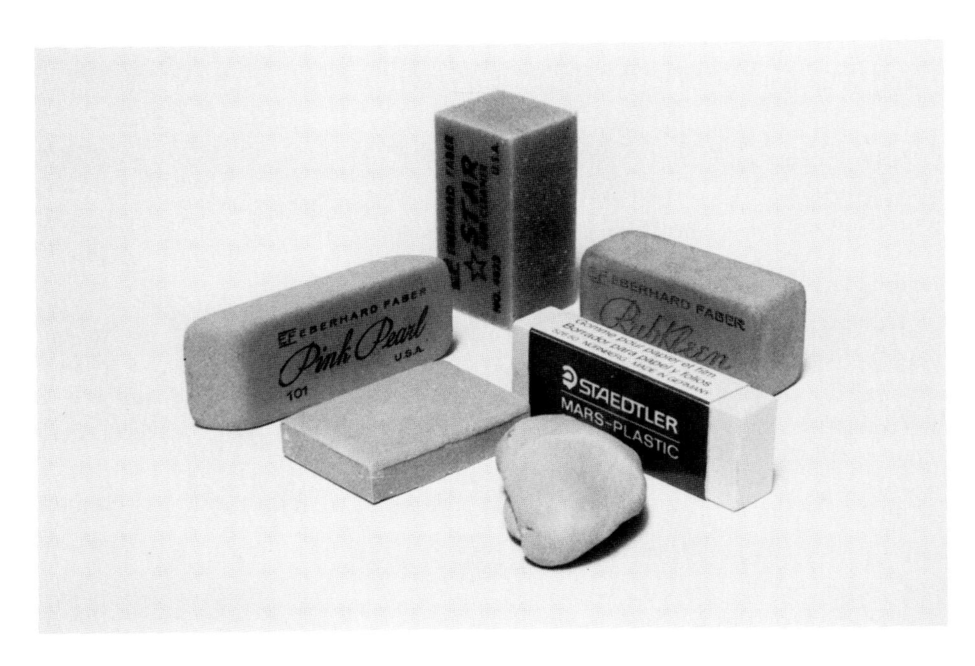

Figure 3.31. Erasers *(from top):* gum eraser (STAR), soft rubber eraser (Pink Pearl and RubKleen), kneaded eraser (Staedtler), ink erasers.

Gum Erasers

A gum eraser (such as Artgum) is the only type of eraser recommended to use on final inked artwork and lettering. It is free of grit and grease and will not mar or scratch delicate artwork. It also crumbles and wears away so as not to damage or lighten original ink and pencil drawings.

Kneaded Erasers

Kneaded erasers are used primarily to erase pencil, pastels, and charcoal and have a soft, pliable, nonabrasive texture. They can be molded into any shape, which lets you use them to erase fine lines as well as large areas. The best feature of the kneaded eraser is that it does not leave particles that get in the way of the pencil or pen strokes and create smearing problems when brushed from a surface. Another feature of the kneaded eraser is that you can press it over dark pencil lines so as to lighten some areas without harming others. When one portion of the eraser becomes blackened with dirt or graphite, knead it to expose a cleaner portion.

Plastic Erasers

Plastic erasers are used to erase ink from various surfaces. The Mars-Plastic drawing eraser is an excellent abrasive-free, vinyl eraser that can be used to remove ink lines from drafting film and paper. An ink eraser made by Pelikan is treated with erasing fluid for easy erasure of inked lines from polyester films with emulsion surfaces and tracing papers.

Electric Erasers

Electric erasers provide neat, quick pinpoint accuracy with only fingertip pressure. To use an electric eraser, insert a round eraser into the tip of the unit and lock it into place; the erasers can be adjusted and/or removed easily (Fig. 3.32). Electric erasers enable you to erase areas quickly and effortlessly. Special attachments can also be used with the electric erasing units, such as a pencil-pointer attachment and a power burnishing head.

Pens

Pens are used most frequently in the comprehensive and advanced comprehensive stage of a design, because of the time required to use them and the level of finish they produce.

Pens are used mainly for finished lettering and the rendering of artwork such as illustrations and logos. Lettering and drawing pens are available with fine points for thin strokes and wide points for broad strokes (Fig. 3.33). This range makes all types of lettering and artwork-rendering possible.

Lettering and Drawing Pens

Graphic artists use immersion-type drawing and lettering pens for a variety of purposes, including lettering, line illustration, and logos. These types of lettering pens are made in two pieces, the handle or ferrule, which holds the pen nib, and the pen nib itself. To use an immersion pen, put the nib, or pen point, in drawing ink so that the ink collects in the reservoir area of the nib. The ink is stored in the reservoir until it is released by pressure.

In order to start the flow of ink, hold the pen at about a 45-degree angle to the paper, as you would hold a pencil. Press the pen tip lightly onto the paper (or other drawing surface) to make your first line. If the line is too thin, apply more pressure; if the line is too thick, apply less

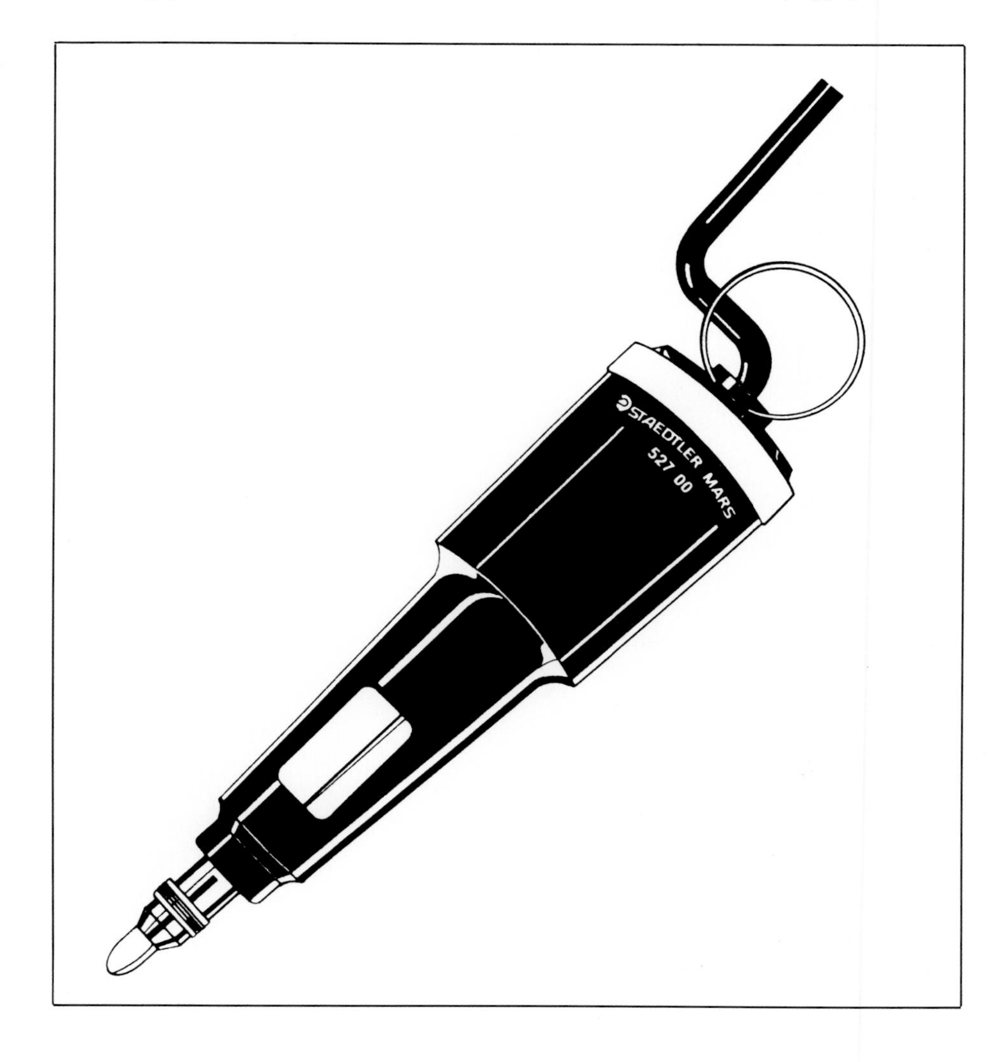

Figure 3.32. Staedtler/Mars electric eraser.

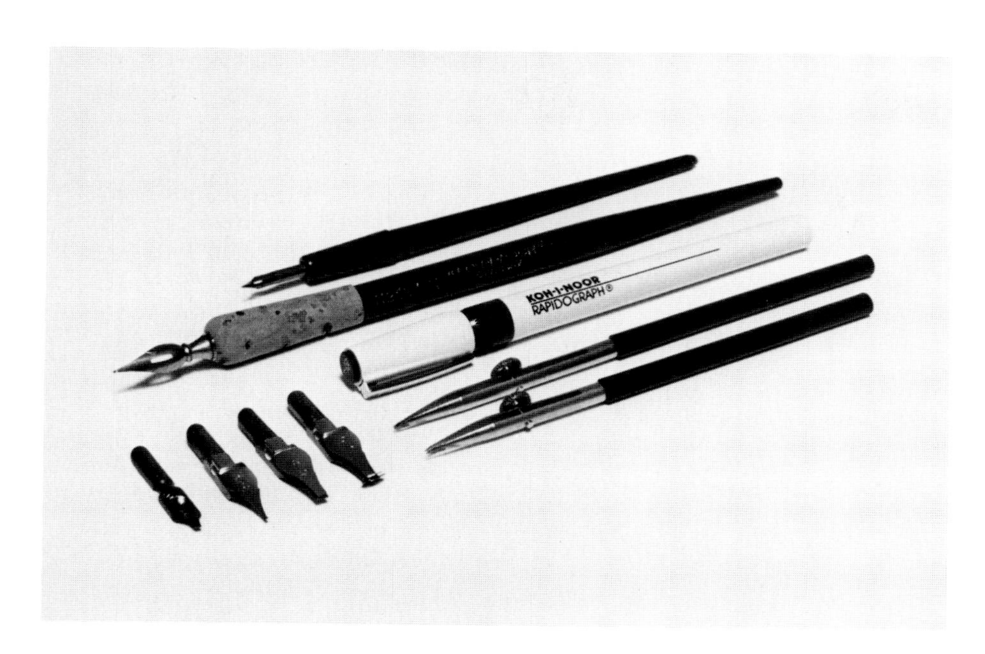

Figure 3.33. Pens *(top to bottom):* crown quill pen, drawing and lettering pen-tip holder, technical pen, ruling pens, and assorted Speedball nibs.

Figure 3.34. Practice strokes made with Speedball C-type nib.

pressure. By changing the pressure on the nib or the direction of the stroke, you can produce a variety of lines and letters. A beginner should first experiment on a scrap piece of paper to become comfortable with how each pen operates.

Lettering and drawing pens are available with different degrees of flexibility and ink-storage capacities. In addition, there are many types of pen holders and different shapes and sizes of nibs.

Nibs are designed to perform different tasks in the design process. Fine-point steel pen nibs, such as the Gillot 170 and the Hunt 22B, are excellent for clean, uniform lines, are easy to clean, and are inexpensive. Speedball nibs include the A type, which has a flat square head, the B type, which has a round head, the C type, which has a square chisel-shaped head, and the D type, which has an oval-shaped head (Fig. 3.34). Each of these nib shapes comes in several sizes to accommodate most design needs.

The square A type is used for making thick, even, downward strokes in lettering and for broad marks in drawing. The round B type is used for rendering curves with lines of even thicknesses. These are good for rendering and filling in broad areas of type and artwork. The flat C type is good for rendering letterforms or illustrations that require thick-and-thin curves and thick, straight vertical strokes. The oval D type is used for rendering sans serif letterforms that contain alternating thick-and-thin strokes and for blocking or filling in large areas with ink.

Technical Pens

Technical pens are used to rule long uninterrupted lines, draw curves, render lettering, and draw freehand line art. There are two types of technical pens, the refillable plastic, cartridge-feed pen such as the Koh-I-Noor Rapidograph (Fig. 3.35), the MarsMatic-700, and the Rotring Rapidograph 150; and the combination cartridge/nib-feed pen such as the Pelikan Graphos.

The cartridge-feed pen is available in thirteen nib sizes, providing a wide range suitable for most applications (Fig. 3.36). It is commonly used for ruling, drawing curves (both freehand and with the aid of templates), and freehand line drawing because the line width never varies no matter what direction the pen is moved in. Cartridge-feed technical pens contain a clear plastic ink cartridge that when inverted continuously feeds the cylindrical nib. To start the ink flow when the pen is first used, shake the pen. Inside the nib is a thin wire that has been weighted so that it can move easily and keep the cylinder free from clogging when the pen is shaken.

The ink used with technical pens often depends on the climate, point size, and drawing surface. Koh-I-Noor and Pelikan provide a wide selection of inks that will suit most of your needs. They make fast-drying inks that can be used on film and large nib sizes, as well as slow-drying inks that are suitable for paper and small nib sizes.

In addition to the nibs, there are other products to use with these pens, such as compasses, drafting films, and erasers and pen cleaners.

Combination cartridge/nib-feed pens such as the Pelikan Graphos have sixty interchangeable nibs, which make them invaluable for ruling lines, lettering and calligraphy, freehand drawing, and sketching

Figure 3.35. The Koh-I-Noor Rapidograph.

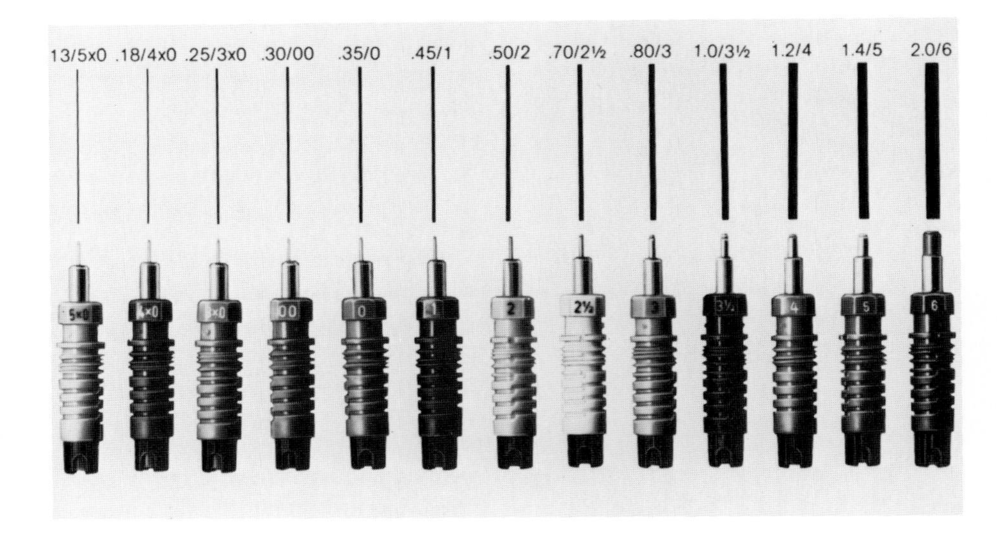

Figure 3.36. Technical pen nibs and their corresponding line-widths.

Figure 3.37. Pelikan Graphos Master Assortment.

(Fig. 3.37). The drawing nibs will not be affected by varying pressure (with the exception of the flexible "S" nibs). The Graphos pen holder houses a cartridge with a visible ink supply. The ink is fed into the nib through a metal tube; the ink flow can be varied by means of an ink-feed regulator. Only ink that is specifically made for technical fountain pens should be used.

In general, the Gillott, Hunt, and Speedball lettering pens and the Pelikan Graphos are best for rendering calligraphy and general freehand finished lettering and artwork because the weights of the letterforms are indicated with only a few strokes. Once the strokes have been made, the technical pen can be used to clean up curves and straighten straight lines. The technical pen is best reserved for ruling lines (use with a straightedge) and for line illustrations that require uniform line weights.

Inks

Inks are available for use in pens and airbrushes and with brushes (Fig. 3.38). It is important to pair the right ink with the right tool for the right surface. For example, if you are rendering with a technical pen on drafting film, use an ink that will not clog the pen point and will not crack or crawl on drafting film.

India Inks

Dense, black India ink is used with most drawing instruments, pens, and brushes for general drawing and lettering on paper, illustration board, and acetate. Which type of India ink you use will depend on the intricacy of the work and the method of application.

Thin, free-flowing Higgins waterproof India ink, which is nonclogging, is recommended for fine-line work with pens, airbrushes, and drawing instruments. Its use, however, is limited to detailed areas because it is too thin to cover large

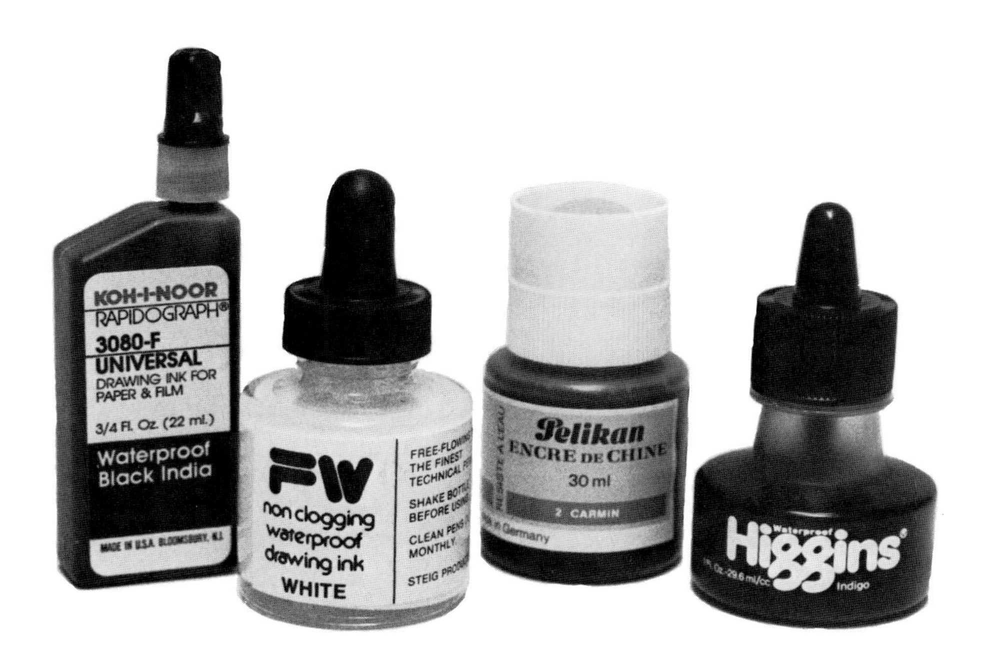

Figure 3.38. Popular brands of inks for use with most art and drafting instruments.

areas efficiently. Higgins Black Magic Ink and Pelikan Drawing Ink can be used to produce both fine-line work and uniform opaque coverage of large areas with brushes or Speedball pens.

Colored Inks

Colored inks are available in opaque and transparent formulas; waterproof colored inks can be used with many drawing instruments, brushes, pens, and airbrushes. Ink colors can be blended or diluted to produce an infinite variety of shades and hues. Red and nonreproduction blue shades are frequently used to prepare mechanicals.

Technical Pen Inks

The choice of ink to use in technical pens will depend on three factors: drawing surface, nib size, and climate. If the drawing surface is a smooth film, such as an acetate or drafting vellum, use a fast-drying ink that will allow you to work quickly without having to wait between strokes for the ink to dry. If the drawing surface is a porous paper or board, use a slower-drying ink that is easily absorbed.

As a general rule, use small nib sizes with slower-drying inks and large nib sizes with faster-drying inks. Climate is related to drying time and thus the choice of ink. Under particularly humid conditions, any ink will require a longer drying time. Koh-I-Noor offers a wide selection of ink, including one called Universal, suitable for most applications.

Color Media

The most popular media for rendering colored elements for use on comprehensive layouts are designer's colors, dyes, acrylics and markers (Fig. 3.39). Which medium you use will depend on a number of factors, such as the size of the area to be covered, budget and time limitations, and the availability of the items.

Designer's Colors

Commonly referred to as gouache, designer's colors are often used to render backgrounds, typography, and small details in the preparation of comprehensives. After thinning the designer's colors to the proper consistency with water, apply them with a brush, ruling or lettering pen, airbrush, or Pre-Val Sprayer. Which tool you use will depend on the size of the area to be covered.

When you wish to achieve a large evenly coated background, thin the gouache to a workable consistency and apply in smooth, even strokes beyond the trim size of the board, using a wide sable or camel-hair brush. In many professional studios, gouache is commonly applied with an airbrush or Pre-Val Sprayer since these are faster methods.

When you want to cover smaller areas using gouache, you can use frisket paper to mask the uncovered areas and then apply the color using a wide brush. Or, you can use fine-tipped brushes, or lettering or ruling pens to fill in small details.

Designer's colors are available in individual tubes and in sets of ten popular colors, including black and white. They are also available in pan form in sets of six, twelve, or twenty-five colors. The basic colors in most sets provide an endless assortment of colors when mixed together (metallic gold and silver can be ordered separately). You can use Winsor & Newton Gum Water, a water-based pale-colored solution, to increase the gloss and transparency of gouache.

Designer's colors are available in a wide variety of shades and hues and can be mixed with one another to produce many more colors. Also, they can be blended on photostats or films to produce gradations or "rainbow" effects.

Dyes

Liquid dyes, which are concentrated and packaged in convenient eyedropper bottles, are available in a wide range of brilliant colors that can be mixed or diluted with water to produce an infinite number of hues and tones. They can be applied with a lettering pen, brush, or ruling pen for small, detailed areas, or with an airbrush for large areas.

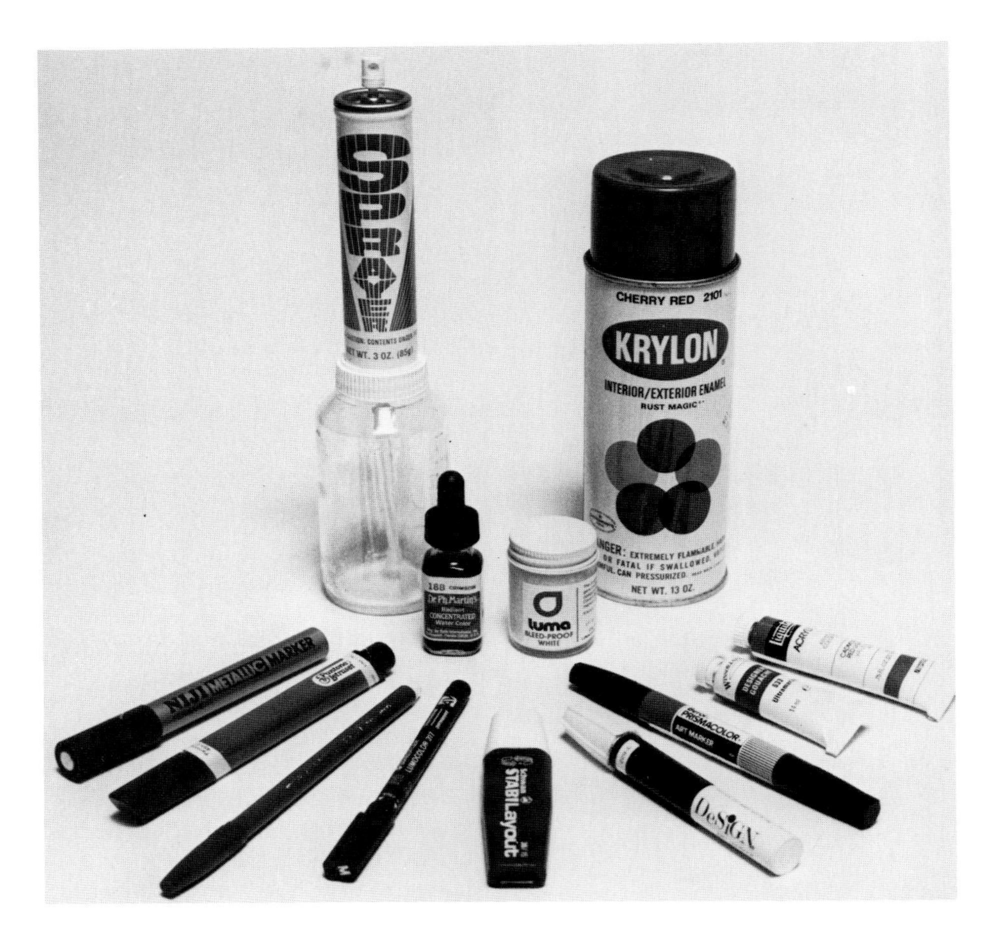

Figure 3.39 *(above).* Color media *(counter-clockwise from top):* Pre-Val Sprayer, Dr. Ph. Martin's Radiant Concentrated Water color (dye), NIJI Metallic Marker, Pantone marker, Marvy Marker, Staedtler Lumo color waterproof marker, STABIlayout marker, Design Art Marker, Prismacolor Art Marker, designer's color (gouache), acrylics, Luma bleed proof white, Krylon enamel.

Figure 3.40 *(below).* Dr. Martin's Radiant Concentrated Water Color set.

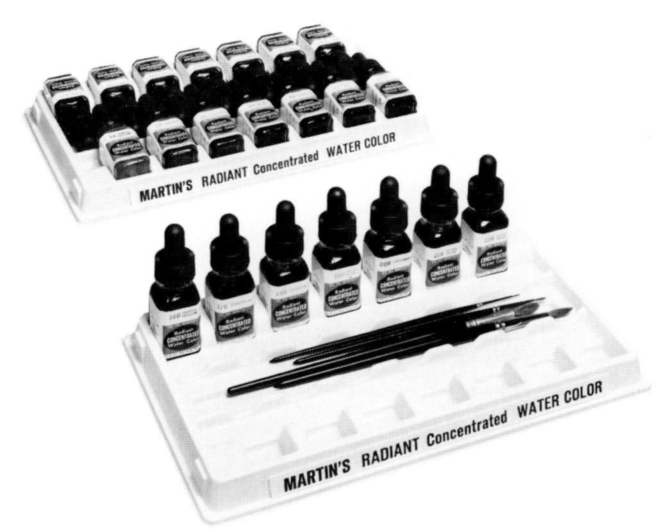

Two dyes are Dr. Ph. Martin's Synchromatic Transparent Water Color and Dr. Ph. Martin's Radiant Concentrated Water Color (Fig. 3.40). Despite the fact that Water Color is clearly written on their labels, they are actually analine dyes that stain whatever paper media they are applied to. The watercolors are available in a wide variety of shades and hues and can be mixed with one another to produce many more colors. Also, they can be blended on photostats or films to produce various color gradations.

Also available from Dr. Ph. Martin are marker tips that screw onto the top of a jar of Dr. Ph. Martin's Water Color. These replaceable, continuous-feed porous fiber tips turn any Dr. Ph. Martin's bottle into a color marker for rendering layouts.

Even though dyes are concentrated, they are very transparent. This quality lets you use them to write over other media (simulating overprinting with ink), hand-color black-and-white photographs, and retouch color photographs. Dyes can also be used as a background and then be covered with gouache. A disadvantage of this technique, however, is that most dye colors will bleed through what is applied over them. Bleedproof white is an opaque white watercolor that is used under dyes, colored markers, and gouache to prevent bleed-through.

Acrylics

Acrylic polymer emulsion paints, or *acrylics* (e.g., Liquitex and Aquatec), differ from water-based media such as gouache and watercolors in that they dry rapidly and become insoluble in water soon after they are applied. This lets you paint over them almost immediately without having the existing color dissolve. After they are applied to a surface, the colors are also durable, flexible, and do not yellow. Acrylics can be cleaned with soap and water.

Acrylics are versatile: They can be thinned and used as transparent washes like watercolors or can be used straight out of the tube and applied thickly like oils. They can also be thinned with water and applied to large areas with a Pre-Val Sprayer.

Markers

Fast, versatile, and widely available, color markers have become one of the most important and commonly used tools in the graphic design studio today. Markers are the ideal medium when time is limited, because they dry quickly, are compatible with other media, and offer a wide color range. In addition, they are convenient to use and require no mixing trays, water bowls, brushes, or general clean-up.

The markers themselves consist of applicator nibs of felt, composition nylon, or synthetic fiber, each available with an extra-fine, fine, medium, broad, wedge-shaped, chisel-shaped, or brush-style tip, and housed in a metal or plastic body. Markers are available in a wide variety of hues, tints, and shades including a complete range of warm and cool grays.

Color markers can be purchased individually or as sets of twelve or more to accommodate most designers' needs. Color charts showing all colors and nib sizes produced by the various manufacturers are available at most art-supply stores.

Color markers come in both spirit- and water-based varieties, each with specific strengths and weaknesses. Waterproof spirit-based markers such as AD Markers, Berol Prismacolor Art Markers, and Design Art Markers bleed through some layout papers but are unmatched in their intensity and brilliance. They can be blended with other spirit-based markers when wet or used with a special Design Art Marker colorless blender that lightens colors and softens sharp edges. The colorless blender can be used for creating a soft single color or multicolor gradations for use on finished layouts, comprehensives, and advanced comprehensives. AD Markers offer interchangeable points, including fine-line, bullet, wedge, and brush shapes.

Water-based markers, such as Marvy markers, STABILayout, and Stabilo-Pen Markers will not bleed through papers but do not have the blending and coverage capacities of many spirit-based markers and often smudge easily. Simple blending of this marker type can be accomplished, however, with a cotton swab or wet brush. The combined use of spirit- and water-based markers on the same layout often produces favorable results. One way of using markers in combination, for example, would be to indicate thin rules or type in a spirit-based marker and render a large background area on top of that. This technique would allow the underlying type to read through the solid background color without the type smearing and would save time by eliminating the need to render the background in and around the type.

Another type of marker is the metallic marker by Pilot, which is available in fine and medium points in silver and gold. These permanent markers will write on virtually any surface, including paper, plastics, metal, foil, and glass. Water-based markers will not adhere to or cover metallic markers, allowing you to cover large background areas with long, uninterrupted strokes. Markers require the least amount of set-up time and maintenance of all the color media. Keep the caps closed on markers when you are not using them and they will last a long time.

Spray Paints

Permanent, fast-drying enamel spray paints are used primarily to render backgrounds and packaging comprehensives. They come in portable, aerosol spray cans that apply an even coat, dry to the touch in minutes on glass, wood, metal, and plastic, require no mixing, and need no clean-up.

Opaquing Whites

Most opaquing fluids are white, but some are red or black. Dr. Ph. Martin's Bleed-Proof White and Pen White are popular brands used for covering or writing over colored backgrounds rendered in watercolor, markers, dyes, and designer's colors since they do not allow the underlying color to bleed through. Dr. Ph. Martin's Bleed-Proof White is usually applied with a brush and Pen White is applied with 3 × 0 or larger technical pens.

Papers and Pads

As a general rule, there is a paper for each step in the sequence of developing a design idea. In the thumbnail stage, any inexpensive paper will do because what is really important is jotting down ideas in words and in visual form as they come to mind. When generating rough layouts, tracing paper can be used to trace type, logos, photographs, and artwork quickly and accurately onto the layout. If the visual elements have been traced on the top sheet of the pad, the sheet can be removed and used as a layout to follow for the next sketch.

As the rough layout develops into a finished layout, layout and visualizing paper should be used because the surface is more receptive to color media and it makes for a better presentation. Layout and visualizing paper is available with different surfaces to accept a variety of media. Bienfang offers four papers: Ad-Art, which comes in pad form, is suitable for use with pencil, charcoal, and pastels; Admaster, which is a 100 percent rag-cotton-content paper that responds to all dry media and felt-nib markers; Art-Vel, which is highly receptive to pencil, pastels and markers as well as to pen or brush and ink; and Graphics, which is made especially for markers, which when applied will hold sharp edges with good color saturation without streaking, spreading, or bleeding onto the sheet underneath.

Bond Paper

Fine-grained white bond paper can be used for sketches and layouts done with pencil, ink, and watercolor. Both heavy- and lightweight bond papers are available in pads in sizes ranging from 9 by 12 to 19 by 24 inches. Heavyweight bond, often referred to as ledger bond, is opaque enough to back translucent tissue layouts for presentation purposes and is transparent enough to use on a light box. Lightweight bond is used only for drawing and sketching.

Lightweight Tracing Paper

Lightweight tracing paper is a transparent smooth paper used for thumbnails as well as for tracing. It can be purchased in rolls or pads in various sizes and weights. Pads are generally easier to use and more practical than rolls. One advantage is that you can slip the same-sized sheets under each other for easy registration and uniform presentation.

Vellum

Vellum, or heavyweight tracing paper, is also used for artwork and mechanicals. You can use it with pencil and ink, and it is durable enough to permit pencil erasing and the removal of ink, which can be scratched off with the edge of a single-edge razor blade. English vellum and heavyweight vellum are also used for color breaks and for instructions on mechanicals or artwork. Some vellums such as Canson Vidalon No. 90 or 110 can be hand-fed through photocopying machines.

Layout and Visualizing Paper

Layout and visualizing paper is a translucent white paper used for doing layouts and comps and for tracing simple opaque shapes. Because it is not as transparent as tracing paper, trace intricate halftone photographs or artwork onto tracing paper first and then slide this under the layout and visualizing paper to be used as a guide for tracing on the next layout. Its slightly textured surface responds well to dry media such as pencil, charcoal, and pastels and allows marker colors to retain their sharpness and color saturation without bleed-through. It is available in pads in popular sizes.

Frisket Paper

Frisket paper is a transparent waterproof tissue paper used to mask photographs and artwork before media is applied by brush, hand, or airbrush. Frisket paper can be purchased with or without a self-adhesive rubber-cement backing. It comes in thin and medium weights. To make a stencil, adhere a piece of frisket paper large enough to cover the entire piece of artwork and burnish it down lightly. Use a frisket or an art knife with a #16 blade to cut a stencil shape before applying media.

Color-coated Paper

Available in matte and gloss finishes, color-coated papers are basically thin, white papers that are colored on one side with ink or paint. They come in simulations of numerous printing ink colors so you can use them to select and match colors. Most color-coated paper surfaces will accept common media such as pencils, ink, paints, and transfers, making them especially useful on comprehensives. A popular medium-priced color-coated paper is Color-Aid, which has a silk-screened coated matte surface and is available in 220 coordinated colors. You can purchase individual 18-by-24-inch sheets of Color-Aid paper or 6-by-9-inch sample packets containing all the colors.

Other medium- and lower-priced coated papers include those by Cello-Tak, and Colorcast flint paper, which has a hard, nonabsorbent, mottle-free glossy finish and is used for all types of comprehensives. Swatch books for all brands of color-coated papers are available at most art supply stores.

An expensive coated paper is Pantone by Letraset Color Coated Paper. Letraset also makes uncoated papers that accurately display color printing on uncoated paper. A glossy surface Pantone paper is also available in 162 colors. This paper has a self-adhesive backing that lets you reposition it easily. There are also graduated tone papers, which simulate airbrushed artwork. These papers are available in twenty-four uncoated matte and self-adhesive glossy colors, ranging from light to dark.

Printing Papers

Large paper manufacturers such as Champion and Strathmore supply printers, paper merchants, and artists-supply stores with text and cover-stock papers in different sizes, weights, colors, and textures. Lightweight text papers are used for producing comprehensives of stationery, magazine advertisements, interior pages of booklets, and packaging dummies. Cover-stock papers, which are heavier than text papers, are used for folders, booklet covers, and covering finished mechanical boards for protection.

Metallic Foil Papers

Metallic papers are thin, foil laminated papers available in 20-by-26-inch sheets. They come with a regular or peel-off backing and in many colors, including green, red, blue, copper, gold, and silver.

Metallic foil papers are used as backgrounds when making comprehensives and advanced comprehensives. They are well suited for two-dimensional pieces such as posters as well as three-dimensional pieces such as packages and signage.

Seamless Background Paper

Large-size black, white, and colored papers are available in durable cover-stock weight for use in photography and display work. They are generally sold in 28-by-44-inch sheets and 12-yard-long-by-53 or 107-inch-wide rolls. Seamless rolls are excellent for achieving a plain or graduated tone effect in the background of photographs that use three-dimensional objects, for example, packaging comprehensives, booklets, and folders or props. The purpose of seamless background paper is to eliminate everything in the background in order to highlight what is being photographed. White seamless paper is used to completely silhouette the model or prop being photographed.

Art Boards

Art boards are used for preparing mechanicals and for mounting layouts, comprehensives, advanced comprehensives, and artwork for presentation or protection.

Illustration Boards

Illustration board is made of white, low-glare drawing paper mounted on a rigid board of regular or double thickness. It comes with many surfaces, ranging from smooth to coarse; the ones commonly used for graphic design and illustration are smooth and medium-smooth. The smooth hot-press finish is suitable for pen-and-ink, airbrush, transfer lettering, and mechanicals. The slightly textured cold-press finish is ideal for pencil, airbrush, markers, designer's colors, and acrylics. The double-thickness board is recommended if you intend to cover large areas with water-based media since it can withstand more stress without warping.

Bristol Boards

Bristol board comes as a white sturdy board in sheets or pads in two finishes that accept a variety of media. The medium-finish, or vellum, board has a slightly textured surface suitable for a wide range of techniques using both wet and dry media. The high, or plate, finish has a smooth surface and is made for pen and pencil artwork and lettering because it will hold delicate lines.

Bristol board is available in thicknesses of from one to five ply, determined by the layers of paper that are laminated together. Two-ply bristol board is used for most drawing, inking, and light-box tracing applications; four- and five-ply boards are used for folders and packaging dummies.

Strathmore has a 500 series of bristol board, made of 100 percent cotton fiber, which takes repeated reworking and pencil erasures without feathering. This series is available in both medium and high surfaces and is available in one- through five-ply sheets.

Chipboards

Chipboard is an inexpensive, rigid, gray board used for backing artwork, constructing models, boxes, and easels, and as a cutting surface on drawing boards and in work areas. It is available in 30-by-40- and 40-by-60-inch sheets of thicknesses ranging from 1/16 to 3/16 inch.

Fome-Cor Boards

Made from polystyrene foam sandwiched between two sheets of white coated paper, Fome-Cor is a lightweight board used for displays, exhibits, and mounting photographs and layouts for presentation. It is easily cut with any blade yet is quite strong and resists warping.

Mat Boards

Mat boards are used for matting and mounting photographs, photostats, and comprehensives. Only smooth-surfaced black, white, or gray boards should be used for matting graphic design and photography work. Colors, fabrics, and textures tend to be distracting elements that can take away from the impact of the piece. The mat should always be an invisible part of a presentation since the piece will have to stand on its own after presentation.

Mounting Boards

Mounting boards are generally used for paste-up work and for mounting or backing artwork and photographs. They are available in sheets of various sizes and thicknesses, although the thicker sheets are used primarily for mounting and backing, and the thinner sheets are often used for making packaging dummies. There are basically three types of mounting board: inexpensive white mount boards, which are semi-gloss-coated on one side and dull white on the other, with a chipboard center, available in 32-by-40- and 40-by-60-inch sizes and in two weights (the lighter-weight board is used primarily for making packaging dummies and the thicker board for mounting or backing artwork); TV mount boards, which are gray on one side and black on the other and come in 16-ply sheets ranging in size from 15-by-20 to 40-by-60 inches; and museum mounting boards, which are specially made of acid-free 100 percent cotton fiberboard, used for mounting finished artwork and photographs and which come in many colors and in 2- and 4-ply weights. (Because of their thin weight and soft, textured appearance, they also are used to mount 4-by-5 and 8-by-10 inch transparencies for portfolio presentation.)

Acetates and Film Overlays

Whether you are preparing mechanicals or assembling advanced comprehensives for presentation, you can use acetates and films to serve many practical and aesthetic requirements (Fig. 3.41).

Acetates

There are several kinds of acetates. Cellulose acetate is available in a wide range of thicknesses, sizes, and sur-

Figure 3.41. Acetates and film overlays.

faces. The thicknesses most often used by graphic designers are .003mm and .005mm. It can be purchased as single 9-by-12- to 25-by-40-inch sheets, or 9-by-12- to 19-by-24-inch pads or 40-by-36- to 40-by-100-inch rolls in clear, matte (or frosted on one side), or prepared surfaces.

Clear acetate is used for overlays on mechanical boards to hold type and artwork above the baseboard. It can be used for inking and covering mounted artwork or comprehensives for protection, as well as for portfolio presentation. But with the exception of acetate inks and paints, most media will crawl (move out of the desired area of placement), or chip on clear acetate. Adding a plasticizing agent, such as Flex-Opaque or Color Flex, will enable opaque watercolors to adhere permanently to clear acetate even when it is bent.

Matte acetate is used primarily for color separation overlays because it will accept any medium. Prepared acetate is clear plastic that has been treated on both sides with a gelatin layer that accepts inks, dyes, watercolors, tempera, and so on, without having to have anything added to the medium. When making corrections, media should be carefully removed with water so as not to damage the surface coating.

Lightweight Cellophane

Lightweight cellophane is a moisture-proof cellulose acetate film used primarily for protecting flat artwork and packaging comprehensives. It is available in rolls of 20- or 40-inch widths and in thicknesses of No. 88 (.00088mm), No. 150 (.00150mm), and No. 300 (.00300mm). It will not shrink, stretch, or become brittle. Cellophane is less expensive than thicker acetate and can be stretched tighter to eliminate ripples that reflect light and distract from the underlying artwork.

To use lightweight cellophane to protect flat artwork that has been mounted on a stiff mat board or same-sized backing board, stretch pretrimmed acetate around the piece and fasten it to the back with small pieces of tape until the ripples disappear. If the acetate cannot be affixed to the artwork, attach it directly to the back of the window mat. Stretch it tightly, and tape it as described above.

To use cellophane for wrapping or protecting a three-dimensional piece, trim the acetate to the necessary size, as you would wrapping paper. Lay the piece facedown on the trimmed acetate so that the overlapping seam will be placed lengthwise on the reverse side of the piece. Hold the seam tightly together with one hand while you apply acetone with a brush along the entire length of the seam with your other hand. Hold it firmly in this position for a few seconds to ensure a tight bond. The remaining two ends or sides can then be "bonded" one at a time in a similar manner.

Acetone

Acetone is a clear liquid commonly used to produce packaging comprehensives since it creates a strong, transparent bond between overlapping layers of lightweight cellophane. It is also used for stripping type and images off plastic, leather, glass, and other materials so that transfers, paint, and other media can be applied. Apply acetone with a brush, absorbent cotton ball, cotton swab, or cloth. Make sure you wear rubber gloves and avoid any contact with this chemical, since it may cause irritation.

Color Film Overlays

Color film overlays are self-adhesive matte or glossy acetate films with transparent colors printed on them. They are available in a wide range of colors, tints, and gradations, including metallics, enabling designers to add large areas of colors to most surfaces. Color film overlay sheets manufactured by Letraset, Cello-Tak, and Zipatone are commonly used brands available in 20″ × 26″ small and large sheets. You can purchase swatch books for each brand that include samples of the available colors.

Color overlays have a wide variety

of uses on many materials, including comprehensives, mechanicals, slides, and audiovisuals. Although solid black and ruby red overlays are transparent, they photograph as opaque, making them excellent for indicating rules and shapes quickly and accurately on mechanicals. The adhesive on the overlays is heat-resistant, which makes them suitable for use with audiovisual media.

Because of their transparent nature, color films can be used over pencil guidelines, ink drawings, and black-and-white artwork. In addition, they allow you to render design elements directly on matte films with ink, designer's colors, or markers, using underlying pencil or pen markings as guidelines. Films can also be layered to achieve mixtures that simulate overprinting with ink.

Color films can be cut to any size and shape and then applied to another surface. To cut the film, use an art knife, making sure that you apply only slight pressure so as to cut through only the film, not the backing sheet. This makes the film easier to remove. After you remove the film, apply it to the new surface using a slip sheet; press firmly and trim any excess film with an art knife or cutting needle.

When applying full or large sheets, be careful not to create bubbles or wrinkles. The best way is to peel away a small section of film along one entire edge of the sheet, fold back or cut away the exposed backing, and adhere the film directly to the desired surface. Once the edge has been firmly adhered, remove the rest of the backing sheet slowly while smoothing and adhering the film itself.

Shading Film Overlays

Black-and-white, self-adhesive shading film overlays are available in hundreds of variations, including standard dot screens, graduated screens, line tints, and special-effect textures. In addition, you can overlap sheets and combine hand-cut shapes with solids or tones to create unique and unusual patterns quickly without the need for clean-up. The matte surfaces of most shading film overlays also accept ink, which lets you customize designs and experiment even further. Shading film overlays can be accurately reproduced by any photographic or printing process.

Dry-Transfer Sheets

Dry-transfer sheets are made from film made up of pressure-released images and a nonstick, wax-coated backing. The sheets are available with hundreds of contemporary and classic typefaces, symbols, rules, and borders from many manufacturers, including Letraset and Chartpak. The sharply printed images are suitable for use in layouts, comprehensives, and reproduction.

Most dry transfers are available only in black-and-white, but some manufacturers offer them in basic colors and metallics. In addition to purchasing ready-made dry-transfer sheets, you can produce sheets with your customized designs in your studio or order them from manufacturers or suppliers. To produce them in the studio, you can use either 3M I.N.T. or Chromatec imaging processes, both of which require expensive equipment.

Drafting Vellum

Drafting vellum is available with a clear Mylar drawing surface, a matte drawing surface on one side and a glossy one on the other, or a matte surface on both sides. It comes in regular and heavy weights in pads ranging from 9-by-12 to 19-by-24 inches and 20-yard rolls from 24 to 36 inches wide.

As a drawing medium, double-matte-surfaced drafting vellum combines the properties of paper and film: Like paper, you can use pencil, ink, watercolor, markers, transfer lettering, adhesive-backed films, and typewriters with them. Their filmlike qualities offer dimensional stability, tearing strength, transparency, and heat- and age-resistance. This means that over time, your artwork is not likely to stretch or peel and should last indefinitely. Furthermore, you can erase inks with drafting-film erasers, cover them up with correction fluid, or scrape them off with the pointed end of an art knife.

A primary medium used on drafting vellum is ink applied with a technical pen. The vellum's transparent quality offers a main advantage over illustration or bristol board: Using a technical pen, you can trace pencil layouts directly onto the vellum. Once you have inked the image onto the surface of the vellum, you can photostat it to any size for paste-up or presentation. Vellum can also be used as a base for adhesive-backed films: The films are adhered to the vellum, cut into the desired shape with an art knife, and then photostatted for use in paste-up or presentation.

Tapes

Tapes are available in many sizes, colors, and shapes, and for a variety of applications (Fig. 3.42). They can be used for packaging dummies, or removing rub-down type or artwork from advanced comprehensives as well as for more conventional needs. A weighted 2-roll (3-inch core) tape dispenser stocked with both white paper tape and transparent tape will serve most needs in any design studio.

Masking Tape

Masking tape is a sturdy crepe-paper tape that conforms to smooth and textured surfaces. Its most desirable characteristic is that it can be removed easily from most surfaces without causing damage to design work. Before using it, however, experiment by adhering it to and removing it from a sample piece.

Masking tape is used to mask or

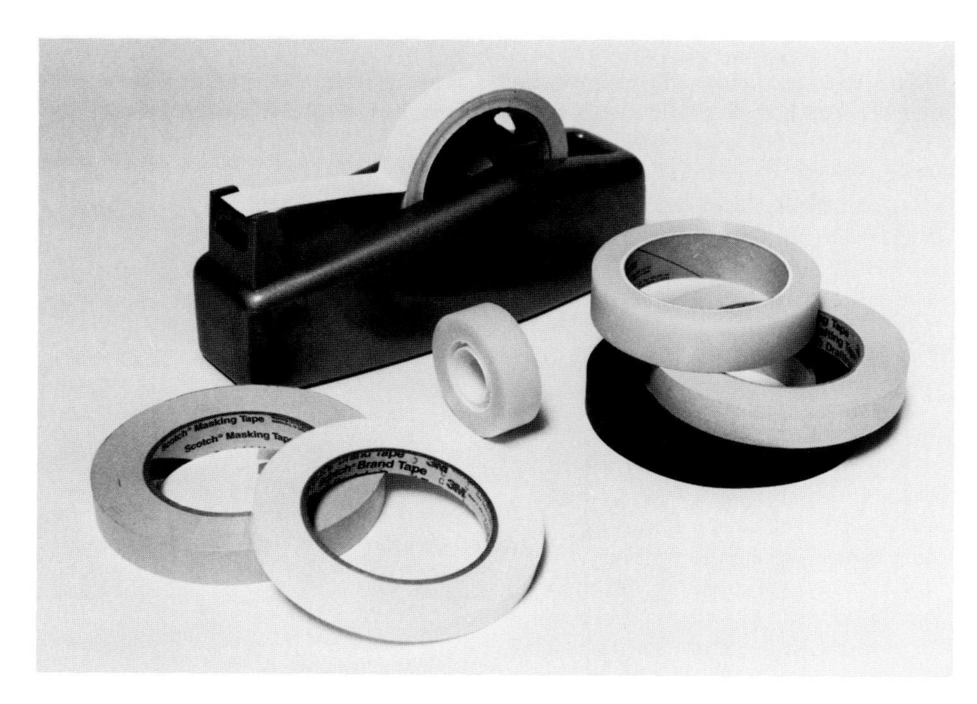

Figure 3.42. Tapes *(left to right):* masking tape, double-sided transparent "Scotch" Brand tape, 3M Magic Transparent Tape, drafting tape, black tape, white tape (in dispenser).

cover up paints and other media, especially overspray from aerosol cans and airbrushes. Other uses include holding work on drawing boards, hinging overlays on mechanicals, bundling and packaging artwork, and assembling set-up box dummies.

Drafting Tape

Drafting tape is a natural colored crepe-paper tape with a low-tack adhesive backing, for holding artwork on drawing boards and other jobs in which tape has to be removed easily without damaging the surface to which it has been applied.

White Paper Tape

Smooth, opaque white paper tape has a low-gloss surface that can be written on easily. Additionally, it adheres firmly to a surface and, as with drafting tape, can be removed easily from most surfaces without damaging them. Its characteristics make it an excellent choice for making corrections and alterations, cropping photographs, and writing instructions on artwork, photostats, and mechanicals. It is also often used for hinging acetate overlays because it blends into the white illustration backing board, allowing clean professional-looking mechanicals.

3M Magic Transparent Tape by Scotch

Manufactured by the 3M Company, Scotch Magic Transparent Tape is a matte-finish acetate tape that becomes invisible when it is applied to a surface. The tape's tough, smooth surface resists cracking and yellowing with age and can be written on with pencil, pen, and permanent markers. It is available in 36- and 72-yard rolls of one-half-, three-quarter-, and 1-inch widths.

Double-Sided Transparent Tape

As its name suggests, double-sided transparent tape is adhesive on both sides, unlike other double-stick tapes that have a liner that must be removed before or during application. This feature, along with its strength and invisibility, makes the tape suitable for packaging dummies.

Black Tape

Black tape, also called photographic tape, is an opaque, black crepe-paper tape used to mask photographs and negative photostats and to hold 4-by-5- or 8-by-10-inch color transparencies in place when they are mounted on black board for presentation.

Patterned Tapes

Patterned tapes are available with solid and broken lines, arrows, dots, register marks, screens, and decorative borders (Fig. 3.43). They range in widths from 1/64 inch to 2 inches and are available in black, metallic tones, and assorted colors. One of the most popular, a precut register-mark tape, is used on mechanical artwork acetate overlays to register the overlay film with the base board. When using adhesive-backed tapes, first cut out the size desired, position it on the artwork with a slip sheet, and burnish it into place.

Adhesives

Adhesives can be applied manually or by machine. The adhesives that can be applied manually, such as glue, rubber cement, and spray adhesive (Fig. 3.44), are less expensive than those that require the use of a machine, such as waxes, and will usually perform as well.

The main advantages to machine-applied adhesives are uniformity and speed of application. Many adhesives are made to perform very specific functions. Common sense usually will tell you which adhesive to use when. For example, rubber cement would be clumsy if applied to the back of a tracing-paper layout that was being mounted on bond paper.

Rubber Cement

There are two types of rubber cement: two-coat, or contact, cement and one-coat cement. Both types create a strong bond between two pieces of most any flat material, and will

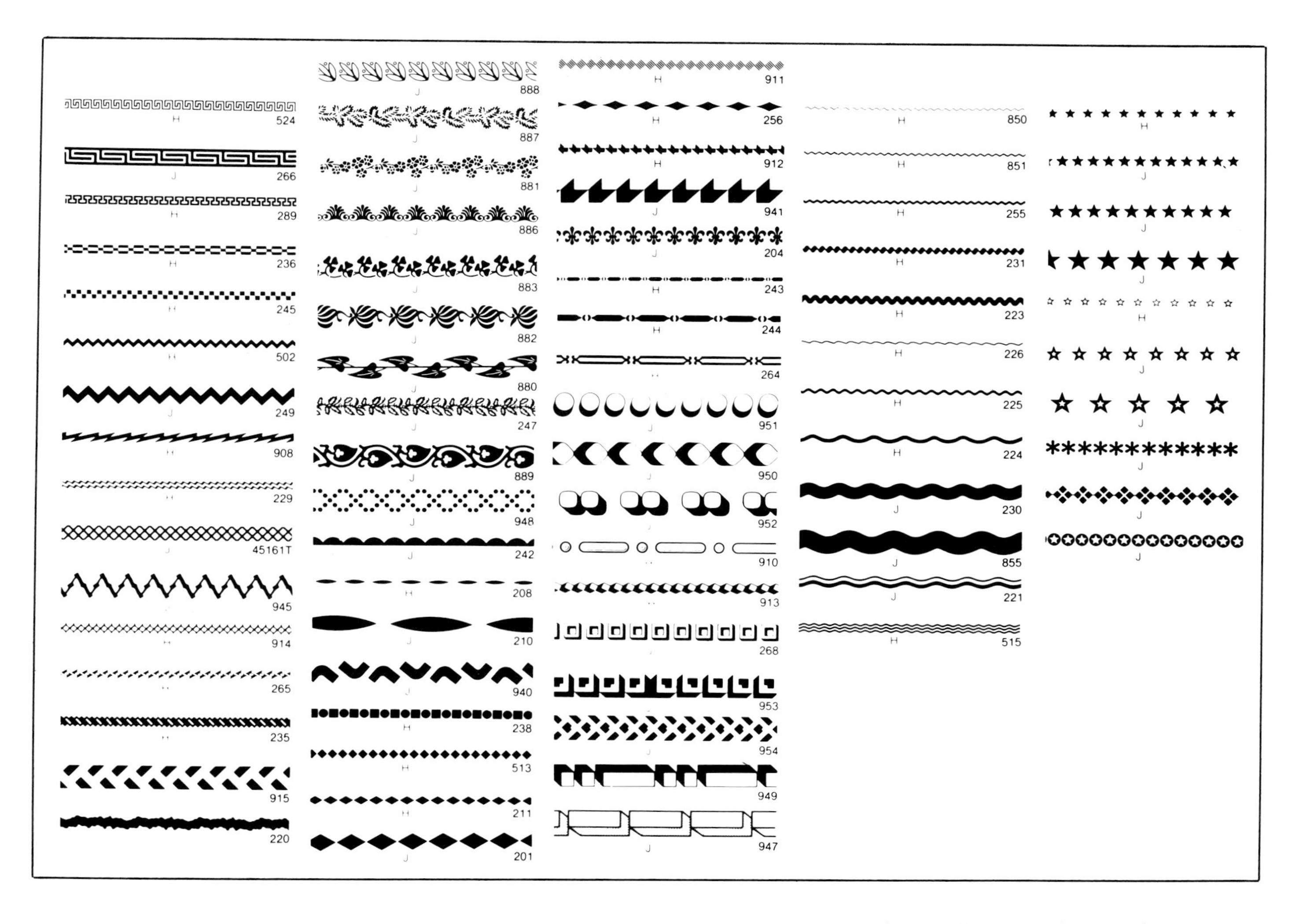

Figure 3.43. Assorted patterned tapes.

Figure 3.44. Manually applied adhesives: *(foreground)* rubber-cement dispenser, thinner dispenser, pickup, Krazy Glue, household cement, white glue, and spray adhesive; *(background)* rubber cement, rubber-cement thinner, and one-coat rubber cement.

not stain, wrinkle, curl, or shrink (even the thin papers), making them especially useful for mounting drawings, sketches, layouts, and photographs. Additionally, they have the advantage of providing different types of bonds for different uses.

For a strong permanent bond, use two-coat cement. Apply a thin but thorough coat to both surfaces to be bonded, allow them to dry, and press them together. For a temporary bond, apply the cement to only one surface and adhere it to the other surface while the first surface is still wet. In either case, remove the excess cement that remains around the edges of the mounted piece with rubber-cement thinner. If you decide to reposition or remove a mounted piece at a later time, simply soak the edge with rubber-cement thinner, allow it to creep under the piece, and remove or reposition the mounted piece before the liquid evaporates.

One-coat cement, which remains tacky for a long time, allowing fast, easy prepositioning without applying additional cement, is applied to only one of the two surfaces that are to bond together.

Although rubber cement performs well for many applications, it does have drawbacks. First, it will not bond any material that is subject to stress, such as warped artwork or photographs, and packaging comprehensives. Second, when rubber cement is used to adhere color-coated paper to a background, the cement may seep through and stain the colored paper. One way to avoid staining is to apply the rubber cement only to the outside trim areas, which can then be covered with a mat. Third, it will eventually deteriorate: Depending on factors such as the method of application, exposure to heat and light, and storage conditions, the bond can last from a few days to several years.

Rubber-Cement Dispenser

Because rubber cement is used in such volume in graphic design studios, it is most economical to buy it in quantity and use it in convenient rubber-cement dispensers. The most popular types of dispensers are airtight glass and plastic jars and cone-shaped metal cans.

The plastic and glass jars are available in pint and half-pint sizes. They have a built-in brush whose handle slides up and down through an airtight opening in the cover, permitting the brush to be adjusted to the level of the rubber cement in the jar so it will not become stiff or dry out. The glass jars have an amber tint; the plastic jars are available in a semi-opaque polyethylene that protects the cement from the damaging effects of sunlight.

The metal-can dispensers have an airtight cone-shaped top that seals a brush inside and prevents the cement from evaporating. The brushes range from 1 to 3 inches in size, making this type of application useful for covering large areas quickly. A built-in wire in the top of the can is useful for wiping excess cement from the brush.

Rubber-Cement Thinner

After a length of time, depending upon storage conditions and exposure to air, rubber cement will thicken to a point at which it may become unusable. When this happens, use rubber-cement thinner to thin the rubber cement to a usable, flowing consistency.

Rubber-cement thinner also has other uses in the graphic design studio. It is an excellent thinner for frisket work, stencils, for loosening mounted material, removing and repositioning artwork, and cleaning design tools and work areas. (Clean your tools and work areas with paper towels saturated with thinner, but work fast because the thinner evaporates rapidly!)

Rubber-Cement Thinner Dispenser

Dispensers for rubber-cement thinner can be made from oil cans and plastic bottles with resealable tops, but the most commonly used dispenser for thinner is the airtight Valvespout dispenser, which is available in a 4-ounce flat shape and a 6-ounce cone shape. Its adjustable brass spout prevents the liquid from evaporating when the dispenser is closed and emits either one drop or a steady stream of thinner when opened. Fill the dispenser by unscrewing the spout at its base and pouring the thinner through a funnel, which will avoid excess spillage. The Valvespout dispenser can also be used for dispensing water into airbrush cups and watercolor palettes.

Rubber-Cement Pickup

Rubber cement pickups are 2-by-2-inch rubber squares used to remove excess rubber cement that has dried around the edges of mounted artwork without damaging the piece. You can use pickups as they are or cut them into small shapes to remove rubber cement in hard-to-reach areas. They can also be used to remove any transfer lettering that has been burnished onto artwork.

Another way to remove dried rubber cement is with the globs of dried rubber cement that accumulate around the rim of the dispenser or with your fingers.

White Glue

White glue is a strong water-based adhesive that does not stain and is transparent when dry. It is used on paper, wood, Styrofoam, cloth, and other porous and semiporous materials. For most uses, white glue is applied directly from a plastic squeeze bottle or tube. Only a small dot of glue is needed if it is applied with a fine-tipped instrument such as a toothpick. When a large area is to be covered, dilute the glue and apply it with a wide, flat nylon brush. For most light materials, spread the white glue thinly on the pieces to be adhered, press the pieces together, and let them dry. For heavier materials such as wood and glass, spread the glue heavily and use a clamp or weight to hold the drying pieces together to ensure a strong bond.

Household Cement

Household cement, also called model-airplane cement, is a super-strength, all-purpose adhesive used for bonding glass, lightweight metal, most plastics, cork, vinyl, leather, wood, and cloth. It dries rapidly and provides a permanent, water-resistant, transparent bond. The most commonly used brands are Scotch Super Strength Adhesive and Duco Cement.

Super Glue

Super glue is fast-acting, transparent when dry, and strong. One drop has thousands of pounds of holding power. It bonds plastic, rubber, metal, glass, ceramics, and many other materials in seconds. A popular brand is Permabond Super Glue.

Spray Adhesive

Packaged in an aerosol can, spray adhesive is a fast-to-apply, clear adhesive that will bond most light materials such as tracing paper, vellum, and newsprint without staining, wrinkling, or bleeding. It will also bond foil, cloth, acetate, and cardboard.

Spray adhesive can be used for a temporary or permanent bond, depending on the way it is applied. For permanent mounting, for example, apply generous coats of adhesive repeatedly to the back of the piece that is being mounted, making sure to let each coat dry. After all coats have dried, press the piece firmly into position on a backing board. For a temporary bond, lightly coat the back of the piece that is being mounted and press the piece firmly into position on a backing board before it dries.

The most popular brands of spray adhesives are 3M Spray Mount, Zipatone Spray Adhesive, and 3M Photo Mount, which is used to mount black-and-white photographs permanently without damaging them.

Use spray adhesives only in areas that are well ventilated to prevent vapor build-up caused by airborne particles. In addition, cover surrounding surfaces adequately and have a large sheet of paper handy to catch any overspray.

Wax

Wax provides a fast, clean, efficient way to adhere paper, photostats, photographs, films, and plastic. After a piece has been wax-coated and pressed into place, it can be lifted off and repositioned without rewaxing. When the piece is correctly positioned, it can be burnished into place for a long-lasting bond.

Wax is applied with electrically heated wax-coating machines that give an even coating without bleed-through (Fig. 3.45). They are available as portable hand-held and large countertop models. With the portable models, you can apply a wax coating that is 1½ inches wide and any reasonable length. With the countertop models, you can apply a coating of up to 12½ inches wide and any length.

Wax is most often used for adhering photostats and type galleys onto mechanical boards or overlays. Any wax residue can be removed with rubber-cement thinner.

Dry-mounting Tissues and Other Materials

Dry-mounting tissues are thin tissues coated on both sides with a heat-sensitive adhesive that is activated when heated under pressure with an iron or dry-mounting press (Fig. 3.46) to provide a permanent bond. Dry-mounting is most commonly used for adhering photographs, but it is also used for bonding smooth papers, posters, lithographs, and even newspaper clippings without resulting in bubbling or shrinking.

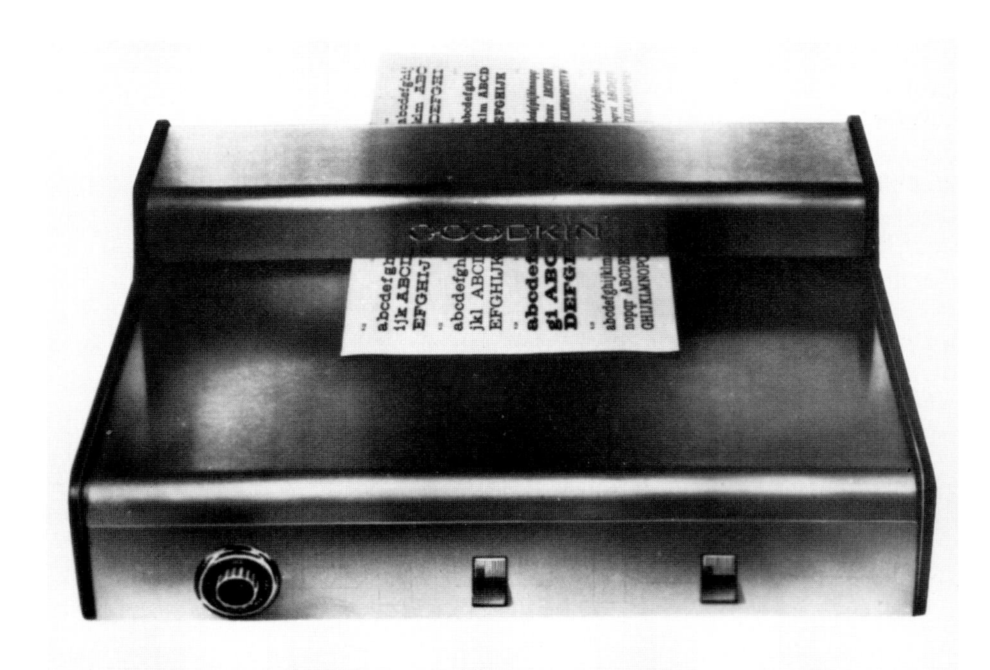

Figure 3.45. Goodkin wax-coating machine.

Figure 3.46. Seal dry-mounting press.

In addition to general-purpose dry-mounting tissues, other types of tissues can be used for dry-mounting. One type is a removable dry-mounting tissue that is activated at a low temperature and is suitable for mounting delicate prints such as watercolors. Another type provides a laminated or clear plastic protective finish (used for drawings and prints, but not photographs). In addition to the dry-mounting press, other tools used with the tissues include a tacking iron, which tacks an adhesive sheet to the back of the piece being mounted; a reusable, nonstick, protective release paper, which is used as a cover sheet over the artwork during dry-mounting; and flat weights, which are used to put even pressure on a piece while it cools.

The dry-mounting procedure is as follows:

1. Gather materials, tools, and equipment, including:
   Mounting press
   Tacking iron
   Weights
   Medium-weight brown wrapping paper
   Dry-mounting tissue
   Release paper
   Piece to be mounted
   Backing board

2. Turn on the mounting press and let it warm up to the recommended temperature for the particular tissue you are using.

3. Dry the piece to be mounted and the backing board separately beforehand to remove any moisture that might interfere with the bonding.

   When the mounting press is at the correct temperature, put a piece of medium-weight brown wrapping paper on the press's sponge pad or base plate. Then put the piece that you are mounting facedown on the wrapping paper, cover it with a sheet of release paper, and close the press and put it in its locked position. The time for drying is determined by the thickness, density, and size of the piece. As a general rule, the thicker the piece, the longer the time required to dry it out.

4. After the piece has been dried, place it facedown on a clean surface, tack one edge of dry-mounting tissue to the back, and trim the tissue to size.

5. Position the piece right-side up on the backing board, lift the edge opposite the edge that has already been adhered, and tack the tissue to the mounting board.

6. Put the piece-tissue-backing board into the mounting press for a sufficient amount of time.

7. Remove the piece and put flat weights over the artwork while it cools.

3M Positionable Mounting Adhesive

Labeled as a cold-mounting technique because the adhesive is not heat-activated, 3M Positionable Mounting Adhesive (PMA) by Scotch is used to perform all the tasks that rubber cement, wax coaters, and spray adhesives perform but with virtually no clean-up required (Fig. 3.47). The adhesive resists changes in temperature and humidity and does not stain, discolor, or dry out with age.

To use PMA, position the piece being mounted and apply pressure in one of two ways to transfer the adhesive and permanent bond. To apply pressure, you can use the 3M plastic squeegee, which is included with all PMA rolls, to transfer and burnish the adhesive onto mounting materials, or you can use a manual two-roller press. The press can handle pieces up to 20 inches wide and ¼

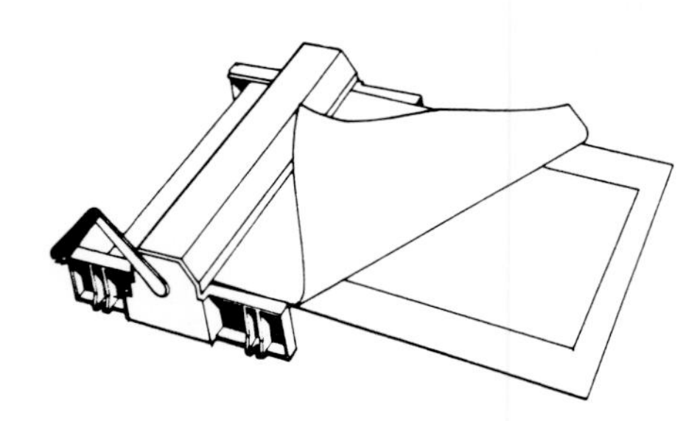

Figure 3.47. "Scotch" Brand 3M PMA applicator.

inch thick without any adjustments needed. After you have burnished the PMA onto the back of the type or artwork, peel away the nonstick backing sheet to uncover the adhesive coating on the back of the type or art (Fig. 3.48). The main drawback of PMA is that it is expensive.

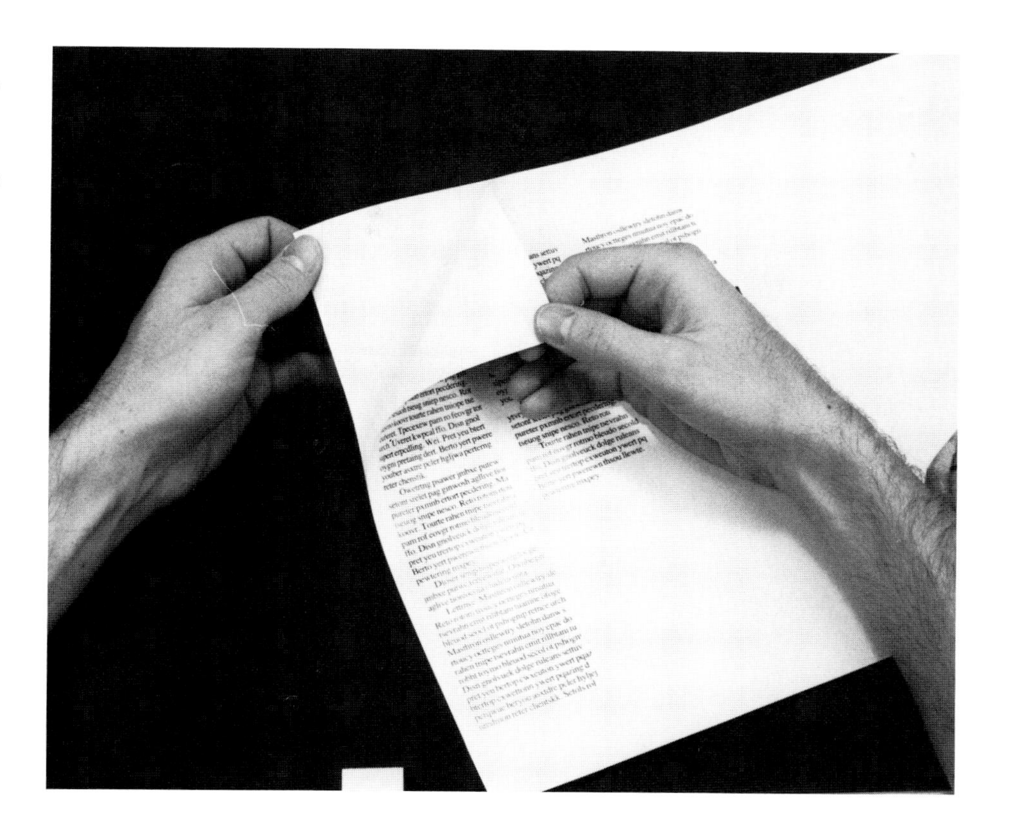

Figure 3.48. Peeling away the nonstick backing sheet to uncover the adhesive coating on the back of the type.

Spray Coatings and Fixatives

Spray coatings and fixatives are used at all stages of the design process, primarily to protect and enhance work.

Clear Acrylic Sprays

Protective clear acrylic sprays, such as Krylon Crystal Clear, provide a permanent, waterproof finish that prevents smudging, fading, and the collection of fingerprints on paper, photographs, type proofs, artwork, plastic, glass, metal, and many other surfaces (Fig. 3.49). It will not discolor.

Clear acrylic spray is available in either a glossy or a matte finish. The glossy finish is used to add sheen or darken the color of a piece of uncoated paper. It is also used to protect packaging comprehensives that are handled frequently. The matte finish is used to eliminate a glossy sheen and light reflection on photographs, photostats, and coated papers.

Dulling Sprays

A dulling spray is a quick-drying, easily removable spray that eliminates glare and highlights on shiny surfaces. It is used primarily in photography to simplify lighting by eliminating glare on reflective surfaces and thus minimize retouching.

Protecting Sprays

Some sprays, such as Krylon Workable Fixatif Spray, provide a clear protective coating over pencil, charcoal, pastels, chalk, watercolor, and India ink to prevent smudging, soiling, and moisture from ruining artwork. The spray dries in seconds to a

Figure 3.49. Four commonly used acrylic sprays for coating and preserving artwork.

workable matte finish that can be worked on and that will accept additional media as well as allow for easy erasing.

Wipes and Cleaning Aids

Wipes and cleaning aids have many other uses in the design studio besides eliminating dirt or correcting mistakes. They can also be used for such processes as applying developers and removing coatings, and they are available in a wide variety.

Wipes

Webril Wipes are soft, lintless, nonwoven cotton cloths specifically used for applying developers and removing excess color coating from photosensitive products without producing scratches (Fig. 3.50).

Cleaning Aids

Absorbent cotton, cotton swabs, facial tissue, bathroom tissue, and paper towels have many uses in the graphic design studio. Absorbent cotton, which is available as rolls and balls, is used to apply ink and dyes as solid colors or washes, to blend soft pencil graphite, charcoal, and pastel, and to apply rubber cement thinner. Cotton swabs perform functions similar to those of rolls or balls except that they are best suited for small areas. Saturated with rubber-cement thinner, cotton swabs are often used to remove excess rubber cement or dry-transfer residue from comprehensives, for blending color that has been added to white opaque Color Keys, and for wiping off retouching color and ink from photographs and photostats.

Facial and bathroom tissues that have been moistened with rubber-cement thinner can be used for cleaning comprehensives, delicate drawing instruments, and mechanical boards. In an emergency, bathroom tissue can be used to apply developer and remove color coatings from 3M photosensitive products. This is not recommended because it may scratch their surfaces. Use paper towels for heavy-duty jobs, including cleaning tools, desktops, and work areas. Paper towels can also be used to "pat" dry Color Keys that have been developed and rinsed.

Figure 3.50. Webril Wipes are used for applying developers and removing excess color coating from photosensitive products without causing scratches.

Presentation Materials

Once design work has been completed, it should be incorporated into a uniform, organized format for presentation. When a body of design pieces has been assembled together, the work can be evaluated more easily by others. A suitable presentation vehicle is a necessity.

Multi-Ring Presentation Books

Multi-ring presentation books are lightweight, portable books used to present layouts, photographs, comprehensives, and two-dimensional printed pieces (Fig. 3.51). The cases are available in real or simulated leather finishes and contain clear acetate sleeves into which the layouts, photographs, and so on, are slipped. The sleeves range in size from 8½ by 11 to 18 by 24 inches. Pieces can be inserted vertically or horizontally, depending on their sizes and configurations.

Multi-Ring Presentation Cases

Multi-ring presentation cases are similar to multi-ring books except that they have a zipper that seals the entire case and a carrying handle (see Fig. 3.51). They also generally have pockets for carrying transparencies, slides, and bulky pieces.

Some zippered cases do not have pockets or ring binders. These are suitable for storing and carrying pieces, but not for presentations.

Rigid Presentation Cases

A rigid presentation case is a deep attaché-style box that comes with or without inside compartments (Fig. 3.52). Its advantages over other types of presentation vehicles are its extra protection and the fact that you can use it to present both two- and three-dimensional pieces. The disadvantages of rigid presentation cases are that they are heavy and it may be difficult to keep the pieces in order if they are viewed by several people.

Three-Ring Binders

Notebook-style three-ring binders can be used to store layouts, slides, reference materials, photocopies of tissue layouts, and job descriptions. In addition, they can hold slides stored in vinyl or plastic prepunched sleeves.

Figure 3.51. Presentation cases *(top to bottom):* multi-ring presentation book, multi-ring presentation case, and two versions of rigid, box-type presentation cases.

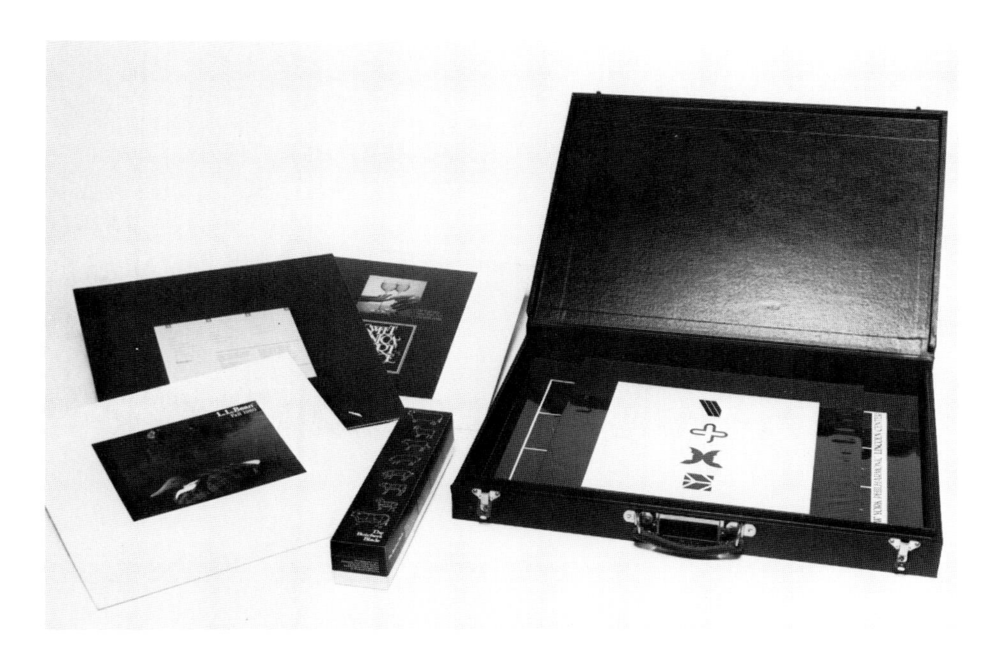

Figure 3.52. A rigid, box-type presentation case can accommodate both two- and three-dimensional pieces.

Basic Techniques in the Graphic Design Process

In Chapters 2 and 3 you learned that when you prepare thumbnails, layouts, and comprehensives, you need to indicate text and display type, render artwork, and transfer designs and layouts from one type of surface to another. The basic graphic-design techniques presented in this chapter explain how to accomplish these and other related tasks. You should begin by mastering some general techniques and procedures.

MAKING A CHISEL POINT

A *chisel point* is a beveled pencil point resembling a household chisel that is created by sharpening flat sketching pencils (see Fig. 4.1).

The procedure for creating chisel points is as follows:

1. If right-handed, hold an industrial razor blade in your right hand between your thumb and index finger and hold a flat sketching pencil in your left hand with the end to be sharpened facing up (Fig. 4.1a). (If left-handed, hold the razor blade and pencil in the opposite hands.)
2. Working away from your body to avoid injury, slowly press the blade into the wood with your right hand while applying pressure behind the blade with your left thumb. Scrape away a half inch of wood (from the front tip) on all sides of the pencil until the lead is exposed (Fig. 4.1b). The best way to carve away the wood is with repeated low, arching motions that expose the lead slowly without making nicks.

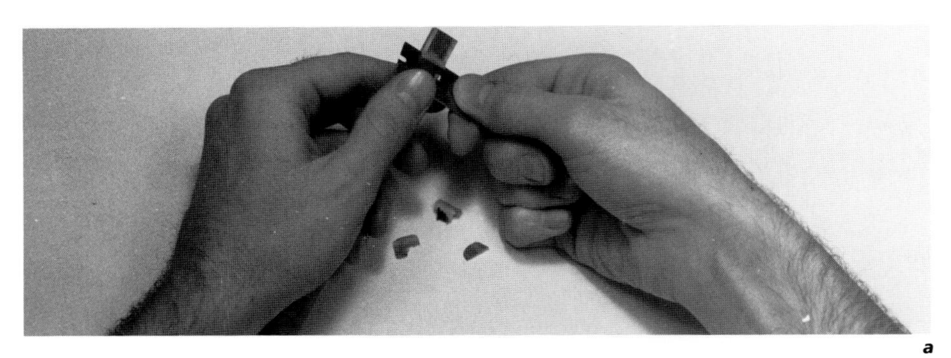

a

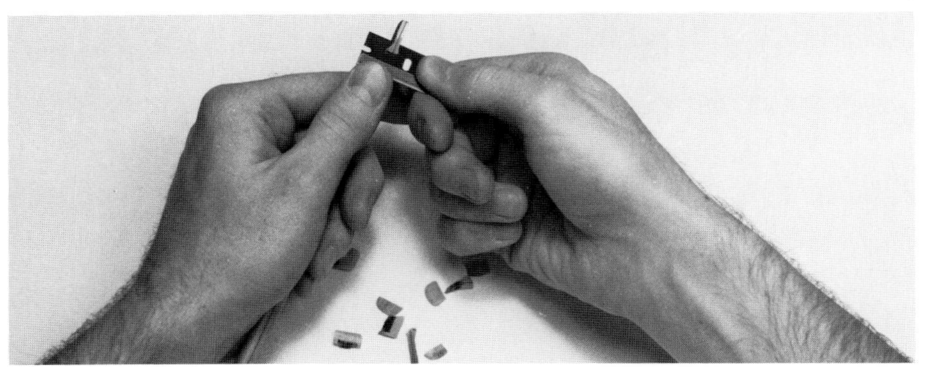

b

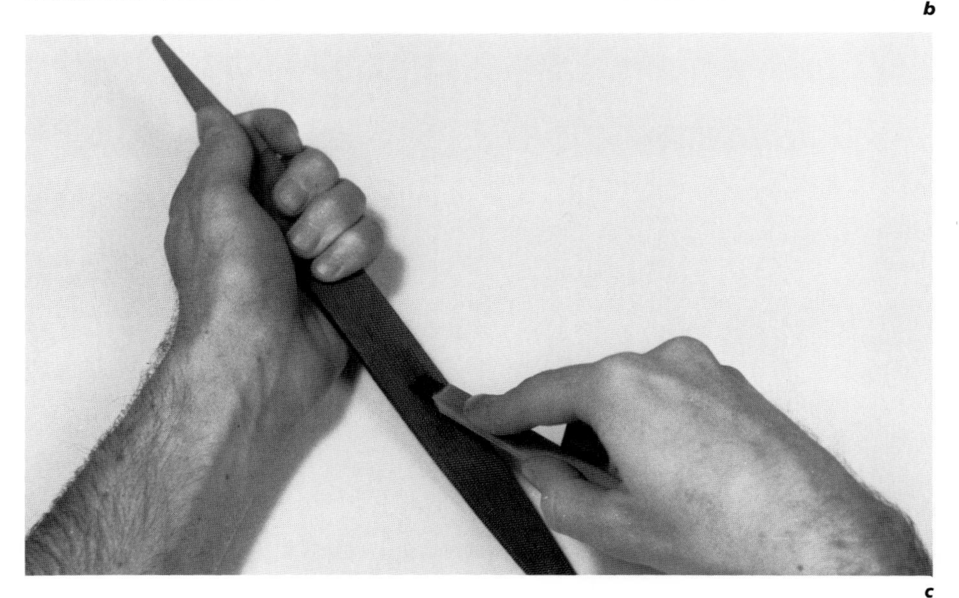

c

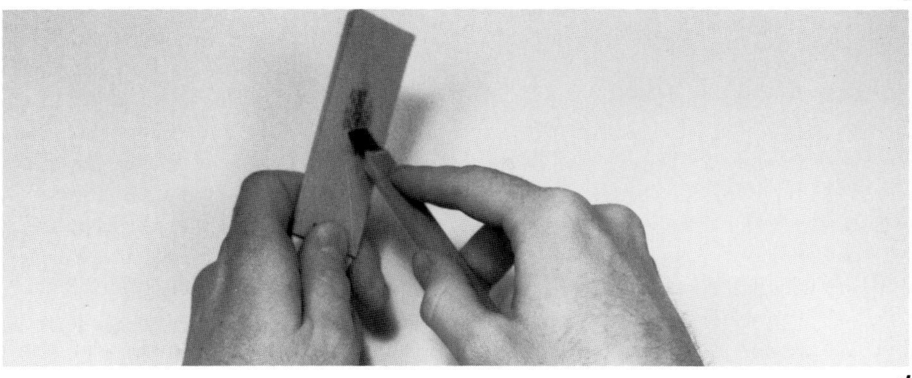

d

3. Now that lead is exposed you can make the point. Hold the pencil in your right hand with your index finger close to the exposed lead. Using a flat bastard file, file the lead on each side until you see two 45-degree angles coming together to form a point when the tip is viewed from the thin side and two parallel lines when it is viewed from the thick side (Fig. 4.1c). This is a chisel point.

Once a chisel point has been formed, you can sharpen its edges on a sanding pad to keep it pointed. When you can no longer sharpen a chisel point, you will have to expose more lead and create a new point (Fig. 4.1d).

Chisel points can also be made on the smaller drawing pencils. Since the lead in these pencils is fragile, use a sandpaper pad rather than a flat bastard file to make the point.

SQUARING UP

To square up is to align your paper, board, or pad horizontally and vertically with a T square (or parallel rule) and a triangle. Once aligned, use masking, or white paper, tape to prevent the item from moving while sketching, drawing, ruling, or preparing comprehensives.

Figure 4.1. Making a chisel point. *a.* Press the industrial razor blade into the flat side of the wood at a low curving arch. In order to remove the wood, it may be necessary to repeat this "scooping" action in order to expose the lead. *b.* Cut and remove the wood from the remaining sides in the same manner without creating any nicks that cause weaknesses in the lead and irregularities in the rendering. *c.* Shape the tip of the lead with a flat bastard file in order to make two 45° angles that join together to form a point. *d.* Once a chisel point has been formed, it can be "repointed" on a sanding pad to sharpen its edges.

MAKING TEMPLATES

As you saw in Chapter 3, commercially produced templates are available in many shapes and sizes. You can, however, also make a template to suit any of your special needs. Templates can be made from many different materials, but the most suitable and popular are varieties of board such as bristol or mat board and 0.015mm-thick clear acetate. Acetate is preferable to board because it is durable and transparent, allowing you to see your artwork through it (Fig. 4.2).

The following are the steps for making a template:

1. Using a pencil, sketch the shape you want onto a sheet of tracing paper.
2. Transfer the sketch onto the acetate or board. If you are using board, follow the graphite transfer method (see Transferring Techniques, page 71).
3. If you are using acetate, lightly spray the back of a sheet of tracing paper with spray adhesive and adhere it to a 0.015mm-thick sheet of acetate.
4. Cut the shape of the template from the board or acetate, discarding the excess material.
5. Lightly sand the cut shape to remove rough edges and smooth out imperfections.
6. If you are planning to use the template for rendering with liquid media, put tape on the underside of the template to prevent it from spreading out or crawling underneath.

Figure 4.2. A homemade acetate template used to render a pattern of curves.

BRUSH RULING

Designers often work on rough surfaces that do not permit them to obtain smooth ruled lines with a pen. In these situations, smooth lines can be obtained with a brush. Brush ruling is often used to create hand-rendered comprehensives, to produce straight lines in all-brush-rendered illustrations, and to indicate lines or shapes when silhouetting photographs (Fig. 4.3).

To render rules with the brush-ruling technique, hold a rigid straight edge, such as an 18-inch wood, steel, or aluminum ruler, in your left hand, with your palm and thumb on top and your other fingers on the bottom. The ruler's top edge should be at a 45-degree angle approximately ½ to 1 inch above the rendering surface, and its bottom edge should be resting on the rendering surface. Hold the brush in your right hand with your thumb and forefinger about 2 inches from the brush tip and your other fingers about another inch back.

To rule lines, place the ferrule of the brush along the ruler's top edge in a position that will allow the tip to make contact with the rendering surface, and literally drag your right hand toward your right. For consistent line widths, maintain even pressure; for varying line widths, simply alter the pressure on the tip of the brush.

DRAWING WITH LIQUID MEDIA

When using a rule or guide to draw lines or shapes with liquid media, you may have to raise the guide slightly above the drawing surface to prevent spreading, or *run-under* (Fig. 4.4). An easy way of raising the guide is to apply one or more layers of tape to the back of the guide.

INDICATING TEXT TYPE

In Chapter 2 you learned that layouts give a general impression rather than a precise rendering of text (or body) type. Specifically, layouts indicate the relative width and height of the different kinds of type to be used, the leading (or spacing) between lines of type, and the location of the type. Techniques for type indication will let you experiment with your layout quickly and easily.

There are two techniques used to indicate text type: the ruled line

Figure 4.3. Brush ruling is often used when producing hand-rendered comprehensives. Lines of varying widths can be created simply by altering the pressure on the tip of the brush.

Figure 4.4. Tape is applied to the back of a drawing guide in order to raise its edge above the paper or board surface. This will prevent spreading, or run-under, of liquid medias.

method and the loop method. The method you choose will depend on the kind of layout you are preparing, the design elements in the layout, the text type, and the tools and media you use. Another consideration when choosing a method of type indication is where the type is being used. For example, if you wanted to indicate copy in charts, where there are rules as well as text, the loop method is better than the rule method.

No matter which method you use to indicate text type, the main requirement is that you make sure the height of the letters is consistent throughout the layout. The guide you use to ensure text type consistency is the *x height* of the text type, that is, the height of the lowercase x in the typeface you choose. Then you begin by drawing *guidelines*.

Text Type Indication Guidelines

Once you determine the type size, weight and length of type to be used in a layout, you are ready to begin. First, square up the paper or pad and outline the area that will contain the text type with light pencil guidelines (Fig. 4.5a). Next, draw a line under the uppermost guideline which is equal to the x height of the typeface being indicated (Fig. 4.5b). These top two lines can be used as a starting point to draw all the guidelines for all the lines to follow. The top line will always be the x-height guideline and the bottom will always be the baseline.

Align a Haberule (along the mark which matches the leading you desire) with the baseline of the first line of type and tick off the lines down to the bottom of the text copy block (Fig. 4.5c). Once these have been indicated, raise the Haberule to the top x-height mark and tick off the measurements down to the bottom of the text block (Fig. 4.5d). Now you have a complete set of guidelines to follow when indicating text type

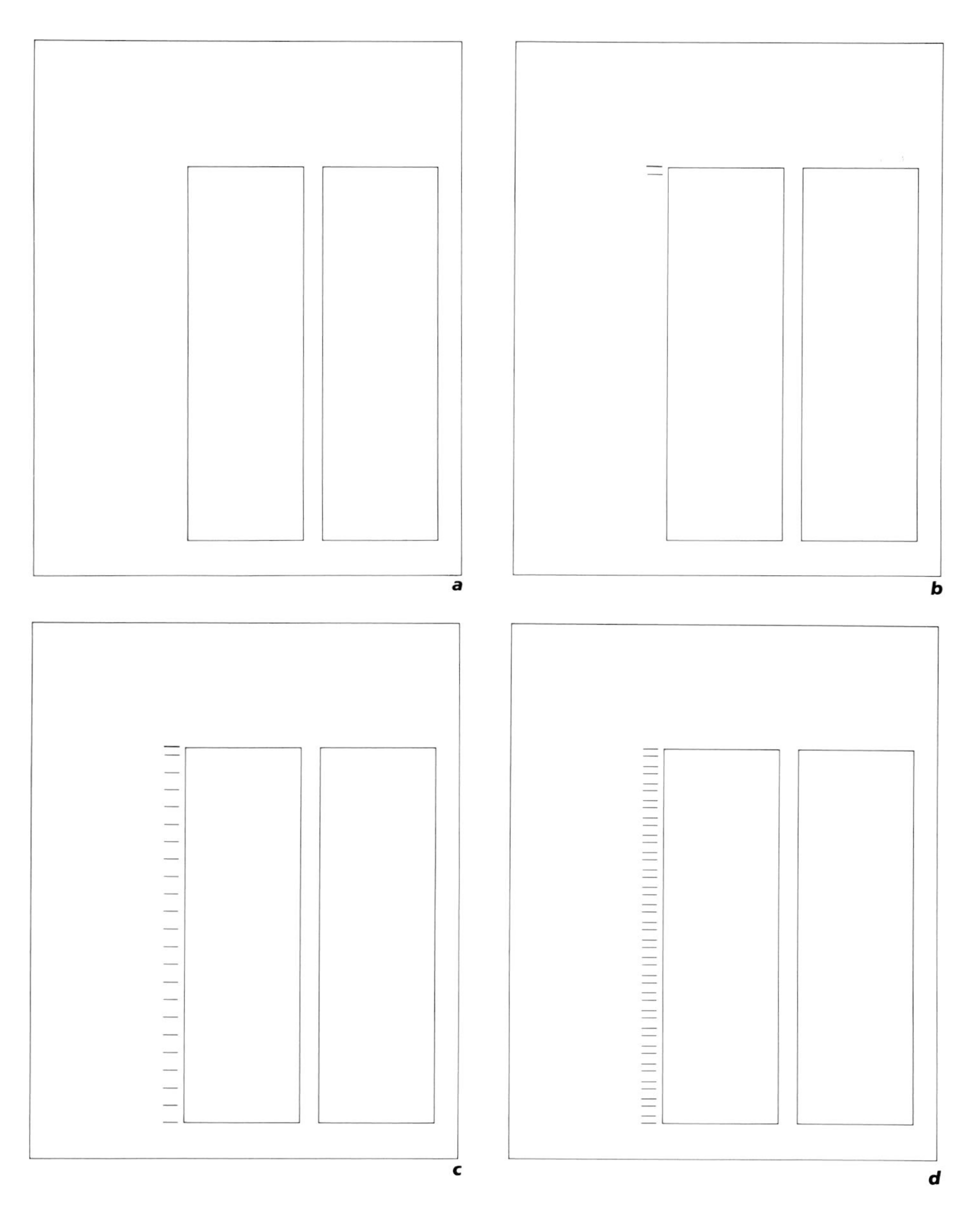

Figure 4.5. Marking a grid. *a.* The outer dimensions of the column length(s) and width(s) should be lightly drawn with a 4H pencil. *b.* Indicate the measurement of the x-height (or cap height if text will be all caps) of the desired typeface on the top left edge of the column and mark its proper baseline measurement below. *c.* Align the point scale of the type gauge that corresponds to the appropriate leading (or baseline measurement) of the typeface to be indicated. Next, starting with the top baseline measurement, mark these off down the entire length of the column. *d.* Shift the same point scale of the type gauge up to the top mark, which is the x-height (or cap height) indication, and mark these measurements down the entire length of the column.

Ruled-Line Method of Indicating Text Type

The fastest way to indicate, or "Greek-in," text type is the *ruled-line* method. You can render ruled lines as one solid line or as two parallel lines; in either case the area covered by the lines indicates the x height of a typeface (Fig. 4.6). The media commonly used for ruled lines range from graphite and ink for gray and black type indication to color markers and designer's colors for color type indication.

When ruling within the x-height area of a typeface, designers must approximate the size and weight of the type and the length of the text. All type is based on the point and pica system of measurement. A point is equal to approximately 1/72 inch; one pica is equal to approximately 1/6 inch. There are 12 points to a pica. All type is specified in point sizes but the point size of type is not its actual size measured in points. Because of the way some typefaces are designed, some point sizes will differ in actual letter size from one to the other. Consult a type specimen book to measure the actual x height. Although actual text type can be as large as 18-point type, you should not indicate it any larger than 14 points. Larger indicated type will not look convincing.

In actual type there is an optical negative and positive (or black-and-white) mixture, or tone, created by many factors, which include the size, weight, and leading of a body of text. For example, if you squint when you look at a block of bold versus regular weight typeset text type, you will notice that they take on different densities. Boldfaced type creates a dark gray optical mixture while lightweight type looks light gray. With practice and experimentation, you will be able to achieve a good representation of this optical mixture when you rule lines to indicate type.

Ruled Pencil Lines

To render ruled lines as one solid line, use a flat sketching or drawing pencil with a chisel point that is the desired x height of the type (see Making a Chisel Point). Using this method, you only need to mark off the baseline measurements (or the horizontal line on which capital and most lowercase letters stand) because the single solid strokes will remain uniform across the column (Fig. 4.7). This method is a fast way to show many columns and pages. After the rules are indicated, spray them with a fixative to prevent smudging.

To render ruled lines as parallel lines, use HB or 2B pencils, which

Figure 4.6. Text type indicated using the ruled-lined method with a 3B drawing pencil that has been cut to the desired x-height.

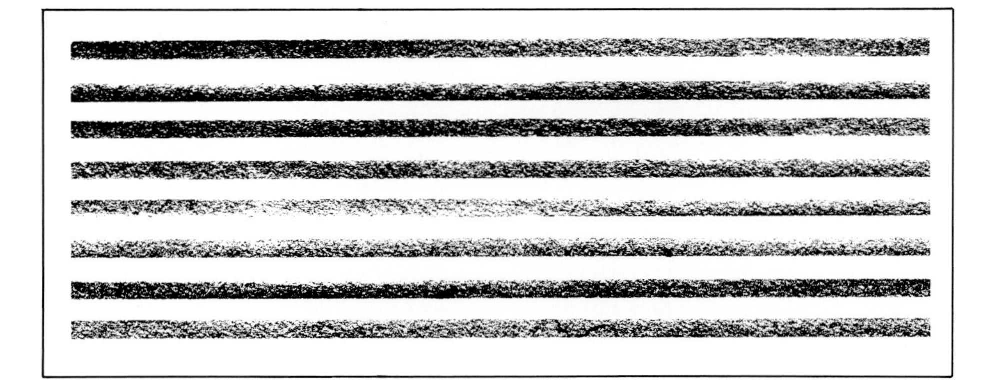

Figure 4.7. Ruled lines indicated with a chisel-point pencil that has been cut to the proper x-height.

do not have to be sharpened as often as softer leads. Because pencil-ruled lines are delicate, draw the underlying guidelines lightly so that they are barely visible and will not require messy erasing later (Fig. 4.8).

To approximate the different weights of the type, you can vary the width of the pencil line rules, provided the height of the rule remains within the x-height measurement. Also, it is important to maintain accurate spacing between parallel lines, flush margins, and proper indentation for paragraphs. As with the solid-line method of text type indication, spray the ruled parallel lines with a fixative before handling.

Ruled Ink Lines

For ruling lines in ink the best tools to use are technical pens, or ruling pens, and a T square or parallel rule. Which pen you use will depend on the kinds of lines you want to show. If your layout contains different sized type that must be indicated by lines of varying widths, use a ruling pen. You can use narrow rules to show the main body text and wider rules to show titles of tables and captions for figures. Ruling pens also can be filled with liquid color media such as dyes or designer's colors to permit fast ruling in any color. If you want to indicate a single-weight line throughout, use a technical pen. You can work faster, and it will not need to be refilled, as do ruling and drawing pens.

When you indicate text type with ink, only use the parallel-rule method since a single solid line of ink does not show the weight of a line of type (Fig. 4.9).

If you look at blocks of type in two very different typefaces, such as a boldfaced Helvetica and a lightweight Garamond, you will note that each has its own texture or color. To achieve a more accurate feeling of text type, you can add texture to ruled ink lines. First, rule the lines with ink and allow them to dry. Then, using a single-edge razor blade, scrape the ink lightly in random areas to produce a textured effect (Fig. 4.10).

Ruled Colored Lines

When a layout calls for ruled lines to be rendered in color, use pointed-nib markers or designer's colors. Colored pointed-nib markers are used primarily for layouts of comprehensives that have been rendered exclusively in that medium. The major drawback to using markers is that they can bleed and crawl on some surfaces, causing undesired lines of varying thickness and tones. To avoid bleeding, test the marker on a small section of the surface you are using to render the

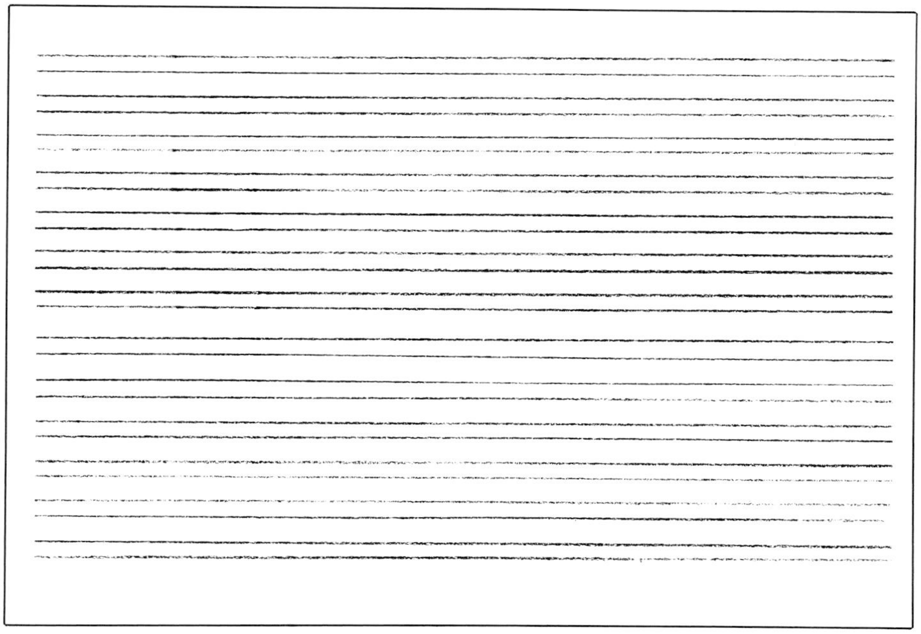

Figure 4.8. Parallel ruled lines made with a 2B pencil sharpened to a fine point.

Figure 4.9. Parallel ruled lines drawn with a technical pen.

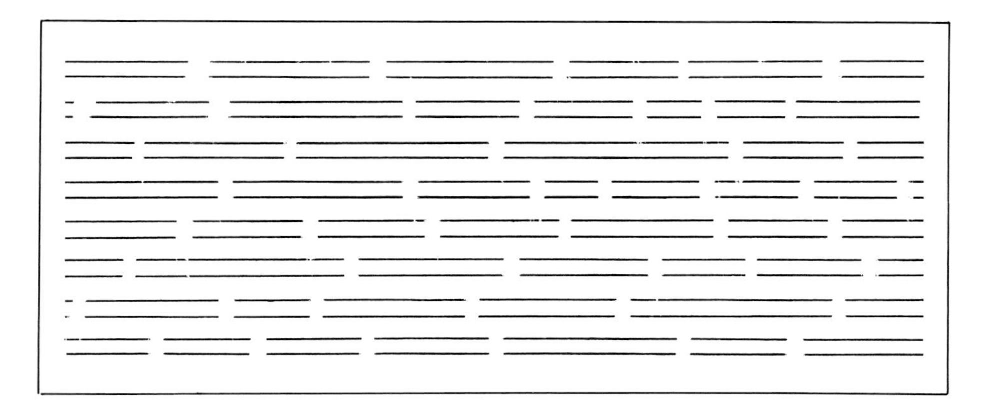

Figure 4.10. Parallel ruled lines, drawn in ink, that have been scratched to more accurately simulate the "texture" of text type.

lines. When using designer's colors to create ruled lines, for best results thin the color with water and apply it with a ruling pen.

Loop Method of Indicating Text Type

The loop method of text type indication is slower than the ruled line method, but it simulates the look and texture of text type more accurately.

Unlike the ruled-line method of type indication, in which you draw one continuous line or two parallel lines to simulate the x height of text, with the loop method you draw loops that are approximately the lengths of words. By varying the length of the loops and the lengths of paragraphs, you will be able to indicate text type quite convincingly. With practice you will learn to produce consistent loops that clearly approximate the characteristics of text type.

With the loop method you can achieve variations in type such as weight and type style. To create a wide stroke or the look of boldfaced type, simply use a graphite pencil with a dull tip, a technical pen, or a marker with a large nib. To render italic type, slant your strokes to the right when drawing loops.

As with the ruled-line method, loops can be rendered in pencil, ink, or markers. The medium you use will depend on the desired effect. If the layout or comprehensive will be rendered in black-and-white, use graphite pencils, technical pens, or markers. If it will be in color, use color markers or colored pencils.

As with the ruled-line method of text type indication, before you indicate text in loops, you must draw guidelines. Follow the technique for drawing guidelines described earlier in this chapter. Just be sure that your guidelines for x height, column width, and column depth are drawn lightly using a 5H pencil. It is important for the guidelines to be barely visible since after you render loops in pencil you will not be able to erase the guidelines.

To speed up the process of indicating text type with loops, put a *type specimen sheet* of the size type and *leading* (the space between lines of type) you want to use under the layout, square it off, and then trace the letters in loop style. You can then draw loops without drawing horizontal guidelines (Fig. 4.11).

When a type specimen sheet is not available, you will need horizontal guidelines so that you can draw loops within the x-height measurement. Strokes above or below the guidelines can be used to indicate uppercase letters and *ascenders* (the part of the type that goes above the line) and *descenders* (the part of the type that goes below the line). By lightly "bouncing" off the T square with the rendering tool on the downstroke of the loop method, you will establish an accurate baseline as well as simulate the rigid, mechanical look of text type.

Figure 4.11. Text type being traced using an underlying type specimen as a guide.

Figure 4.12. Drawing loops using a pencil. *a.* Loops drawn with a HB drawing pencil and a T square, which helps to establish an accurate baseline while also simulating the rigid, mechanical look of text type. *b.* Loops drawn freehand with a HB drawing pencil.

Pencil Loops

To indicate loop text type with pencil, use a graphite HB or 2B pencil. Both have a hard lead, which maintains a sharp point and produces a dark gray texture in lettering. To vary the weight of type when indicating loops, use a rounded tip for boldfaced type and a sharp tip for light, fine type. After the type has been rendered in graphite pencil, make sure you apply a fixative to prevent smudging. Figure 4.12a shows loops that have been rendered with the aid of a straightedge, and Figure 4.12b shows loops drawn freehand.

Ink Loops

The best tool for rendering loops to indicate text type is the technical pen, which produces a consistent line width that closely simulates actual type weight. Technical pens are available in a wide assortment of nib sizes, making it easy to match the weight of the type you are indicating.

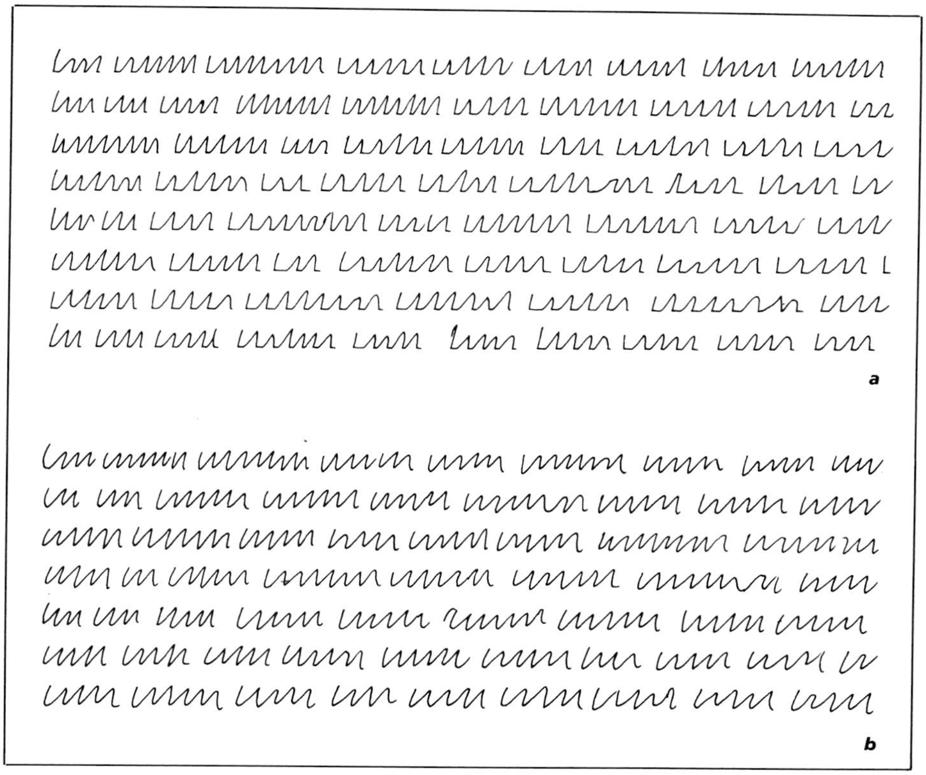

(For example, a 3 by 0 nib will approximate medium-weight type, and a 4 by 0 nib can be used for lightweight typefaces.) Figure 4.13a shows loops drawn with a technical pen using a T square as a baseline guide, and Figure 4.13b shows loops drawn freehand with a pen.

a

b

Figure 4.13. Drawing loops using a pen. *a.* Loops drawn with a 3 x 0 technical pen using a T square as a baseline guide. *b.* Loops drawn freehand with a 3 x 0 technical pen.

Colored Loops

Colored loops can be rendered with colored pencils or markers. Two types of colored pencils are suitable, but each has some drawbacks: Wax-based colored pencils such as Berol Prismacolor give brilliant opaque coverage, but their tips wear down rapidly, making it difficult to render type with consistent line weights. Colored pencils made of hard, thin lead such as Berol Verithin hold a sharp point longer but do not offer the brilliance and opacity needed to render type convincingly in color.

The best medium for rendering loops in color is colored markers, which come in an assortment of colors and nib widths. They offer brilliant colors and the ability to vary type weights simply by changing nib widths. The major disadvantage of markers is that they tend to bleed if you hold the marker in one area for too long. To prevent bleeding, make sure you keep the flow of loops moving.

INDICATING DISPLAY TYPE

In Chapter 2 you learned that display type is used for headings and other material that is set off from the body text. Figure 4.14 illustrates a layout incorporating both text and display type. When you indicate display type on layouts and comprehensives, it is necessary to show its size, weight, and location, as with text type indication. But it is also necessary that the words be readable and the typeface recognizable.

To make sure that you accurately depict a typeface, refer to a type specimen book or type supply catalog (Fig. 4.15). You can either trace the type or draw it freehand. The method you choose will depend upon the amount of time you have and whether or not you can enlarge or reduce the type photomechanically with a photocopying or photostatting machine or a Lucy (see Chapter 3). In either case, make sure you include all of its features, such as its approximate weight, and whether or not it is a sans serif typeface.

Different techniques, tools, and media are used to indicate display type. They depend on the step you are working on in the design process—for example, the type in a comprehensive must be more refined than that in a layout—and the effect you wish to achieve. The following descriptions are of various display-lettering techniques, including the tools and media used with each.

Pencil Lettering

The pencil you use to indicate display type depends on the typeface and size of type you wish to indicate. Ideally, you should render each part of a letter with a pencil cut or sharpened to the appropriate stem width (Fig. 4.16). For example, if you want to indicate a serif typeface such as Bodoni or Caslon, use a chisel-point pencil for the wide curved and straight strokes and a sharply pointed drawing pencil for the serifs and thin strokes, as shown in Figure 4.17. To do this, hold the chisel-point pencil perpendicular to the baseline guideline and maintain that position to indicate downstrokes and curves. Use the sharply pointed pencil to connect the thin strokes and serifs to the wide strokes. If you want to indicate a sans serif typeface such as Helvetica or Futura, use a chisel-point pencil, which lets you create letters of even widths with a series of wide, connected strokes, as shown in Figure

The New York Times

Section 10

TRAVEL

Sunday, March 9, 1980

Figuring Out New Air Fares

Jazz in Europe is Alive And Swinging

Getting Back To The South Pacific

Inside: Guildhalls in London 3 · Influx Into Daytona 5 · Medieval Pérouges 7 · Classes in France 10 · Notes: Goldpanning 11 · What's Doing in Bern 13 · Practical Traveler 19 · Letters: Costa Rica 25

Figure 4.14. A design that incorporates both text and display type rendered with a 3 x 0 technical pen.

Figure 4.15. Type specimen books and type supply catalogs from typesetters are excellent reference sources to use when rendering display type by hand. Here letters are accurately traced with a 3 x 0 technical pen.

Figure 4.16. Chisel-point pencil cut so that its width will indicate the stem of a letter with one stroke.

Figure 4.17. Display type rendered with a chisel-point pencil.

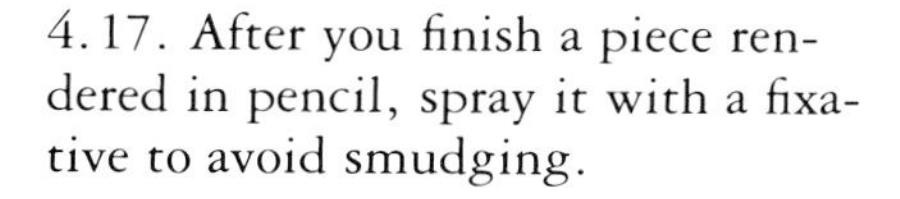
4.17. After you finish a piece rendered in pencil, spray it with a fixative to avoid smudging.

Ink Lettering

Display-type ink lettering can be rendered with a brush and a split-nib drawing or lettering pen, or a technical pen. To render letters in ink, first outline them using the fine nib of a drawing, technical, or lettering pen and then fill them in using a brush or the wide nib of the drawing pen (Fig. 4.18). When using a drawing pen to fill in large areas, fill it with extra-dense ink, which is opaque; for small, detailed areas, fill the pen with extra-fine ink, which will not clog the pen point.

To render fine display type for comprehensives, many designers use technical pens, which give smooth lines of consistent width. You can also use the pens with a T square and triangle to draw vertical, horizontal,

Figure 4.18. Logotype outlined on multipurpose drafting vellum with a 3 x 0 technical pen and filled in with a B-4 Speedball pen.

and angled lines that are parallel and even, leaving only curves to be rendered without the aid of a straightedge.

Color Lettering

Display type can be rendered in color using a variety of media and tools. Type can be indicated with designer's colors applied with a drawing pen, brush, or ruling pen, depending upon the size of the type. To indicate small type or outlines, fill a drawing or ruling pen with a designer color that has been thinned with water to allow it to flow smoothly and cover the area effectively. To render broad strokes or letters, use a pen with gouache, taking care not to let it thicken in the tip. Figure 4.19 shows lettering rendered with designer's colors and a red sable brush.

The tools, media, and technique used to render display type are determined by the background surface on

Figure 4.19 *(left).* Lettering rendered with designer's colors and a red sable brush on colored paper. (Artwork: Paula Warnagiris)

which you draw the letters. When lettering on a smooth, light-colored background, use a technical pen to indicate small details and maintain even strokes. When lettering with gouache over a painted background, a brush is more suitable since it will not clog or scrape off the underlying surface.

Figure 4.20 *(above).* Lettering rendered with fine- and broad-tipped markers on layout and visualizing paper.

Color Marker Lettering

Color markers offer advantages over other media used to letter display type. They come in a variety of nib sizes and shapes to accommodate different types of lettering. There are both spirit- and water-based markers, allowing you to create a background with water-based color markers and then letter over it using spirit-based markers, which will prevent bleeding. Also, since the markers are transparent, you can apply color in layers and simulate *overprinting,* or the printing of one ink or image over another. Color markers are generally used to indicate display type for layouts or comprehensives that have been rendered on layout or visualizing paper (Figure 4.20).

Cutout Lettering

Another way in which to indicate display type on a layout or comprehensive is with letters cut out of various media. The technique of cutout lettering involves cutting out individual

Figure 4.21 *(left).* Lettering that was cut out of paper with an art knife and adhered to colored paper with spray adhesive. (Artwork: Randy Tibbott)

letters or shapes and pasting them onto layout or comprehensive is time-consuming and thus suitable only for large simple letters, words, logos, and so on (see Fig. 4.21).

The media most often used for cut-out letters are Pantone adhesive-backed paper and film, foil, textured paper, fabric, or any material that gives the desired effect can be used. Just make sure it can be cut easily and accurately.

To ensure accuracy when cutting out letters, it is important to use a sharp tool such as an X-Acto art knife. Art knives can be used with any metal straightedge to produce clean horizontal and vertical lines. When cutting letters out of a hard-to-handle material such as fabric, use a scissors to cut the outside of the letterform and an art knife to cut out the inside and difficult-to-reach areas.

To cut letters out of paper with an art knife, proceed as follows:

1. Lightly coat the backside of the sheet of paper from which you will cut the letters with spray adhesive.
2. Adhere the paper to cutting or illustration board to ensure a smooth, firm cutting surface.
3. Cut out all the letters you will need and then peel the paper letters away from the backing board.
4. Apply a heavier coat of adhesive spray or rubber cement to the back of the letters and cement them to your layout.

If the paper from which you cut your letters has an adhesive backing simply cut out the letters and then remove the backing. Attach the letters to your layout and burnish them down firmly.

When cutting letters out of film with an adhesive backing,

1. Adhere the film onto the layout or comprehensive.
2. Cut out all the letters.
3. Peel away the excess film.

RENDERING TYPE ON ACETATE OVERLAYS

When preparing a comprehensive, the type, logos, and other foreground elements are often rendered on a clear acetate overlay rather than on the background paper. Acetate is used when it is not possible to draw on the background medium, as in the case of one done with cut film (Fig. 4.22), or when the medium may bleed, such as dyes, or the surface is too rough, as in the case of acrylics. Prepared acetate is recommended since it accepts almost any medium (see Chapter 3 under Acetates and Film Overlays).

To render type or designs on acetate, first secure your ink or pencil layout to a backing board. Then tape an acetate sheet over your layout, allowing enough excess so that it can be covered later by a mat. It is important to keep the acetate flat over the layout to ensure accuracy when transferring or inking type.

To render small type or to outline large type, use a fine-tipped sable brush; to cover large areas, use broad-tipped brushes. Although ruling pens are difficult to use on acetate, they can be used to rule lines and draw patterns.

After the type or foreground design elements have been rendered, put the overlay over the background, stretch it flat, and secure it in place with tape attached outside of the image area. Finally, drop a window mat over the image area to protect it and prepare it for presentation.

TRANSFERRING TECHNIQUES

As you have seen, the graphic design process is done in steps making it necessary to *transfer,* or duplicate, images from one surface to another. There are several techniques you can use to transfer your layouts from transparent tracing, layout, or visualizing paper to opaque surfaces such as illustration board.

Figure 4.22. Foreground elements (type) are commonly rendered on acetate when the background surface is unsuitable for direct rendering. Here white type is being "painted" onto acetate overlaid on a cut-film background.

Graphite Transfer Method

The most accurate way to transfer an image from a layout on a transparent surface to an opaque one is the graphite transfer method. To accomplish this technique:

1. Using a soft 2B drawing pencil, apply graphite to the back of a tightly rendered tissue layout (Fig. 4.23a). Make sure you only apply the graphite to the image areas that will be transferred. When transferring a tissue layout onto a dark background, use a white pastel pencil instead of a 2B drawing pencil.

2. Put the tissue layout *right reading,* that is, face up, in the desired position on the paper or board on which you will transfer the image (Fig. 4.23b). Tape the layout firmly in place.

3. Using a hard, sharp, colored pencil or fine-tipped color ballpoint pen, trace the image you want to transfer. Make sure you use a color other than the one used on the layout or comprehensive (Fig. 4.23b).

4. Once you have traced the image, remove the lower tape and lift up the tissue layout to verify that the entire image has been transferred. If any lines were not transferred, retape the layout and trace them. This procedure ensures that the lines are where they belong (Fig. 4.23c).

5. Once all the lines have been accurately transferred, remove the tissue layout. You can now render the type and artwork using the appropriate technique (Fig. 4.23d).

a

b

c

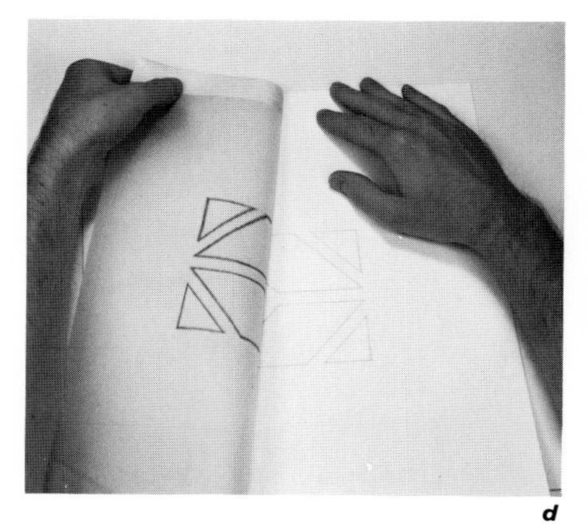

d

Light Box Tracing

A light box (see Chapter 3 under Equipment) is the cleanest, easiest way to transfer type or artwork from one surface to another. To transfer images using a light box, lay the original artwork or type face up on the illuminated light box and place a sheet of paper, lightweight board, or film over it. You can now trace the image using a pencil or the actual color medium used in the layout or comprehensive (see Fig. 4.24).

When using a light box, make sure the materials are translucent enough for light to pass through. For example, most paper, lightweight boards such as one- or two-ply bristol boards, and films permit easy tracing, whereas illustration boards are too heavy and opaque.

ENLARGING THUMBNAIL SKETCHES

When you translate your thumbnail sketches to their reproduction size in layouts, you will have to make some adjustments in the designs to accommodate the exact dimensions of the layout.

Enlarge the thumbnail to reproduction size by one of the following methods: visual estimation, mechanical enlarging such as with a Lucy or opaque projector, or photographic enlarging using a photocopying or a photostat machine.

Figure 4.23. Transferring type. *a.* Apply graphite to the back of a tightly rendered tissue layout with a soft 3B drawing pencil in the image areas only. *b.* Turn the tissue layout over and trace the entire (right-reading) image with a sharp, hard, colored pencil or fine-nib color ballpoint pen. *c.* Check to see if the entire image has been transferred by removing the lower tapes only. This method will allow for easier repositioning later if any lines were not transferred. *d.* Remove the tissue layout. The remaining image can now be rendered following these markings.

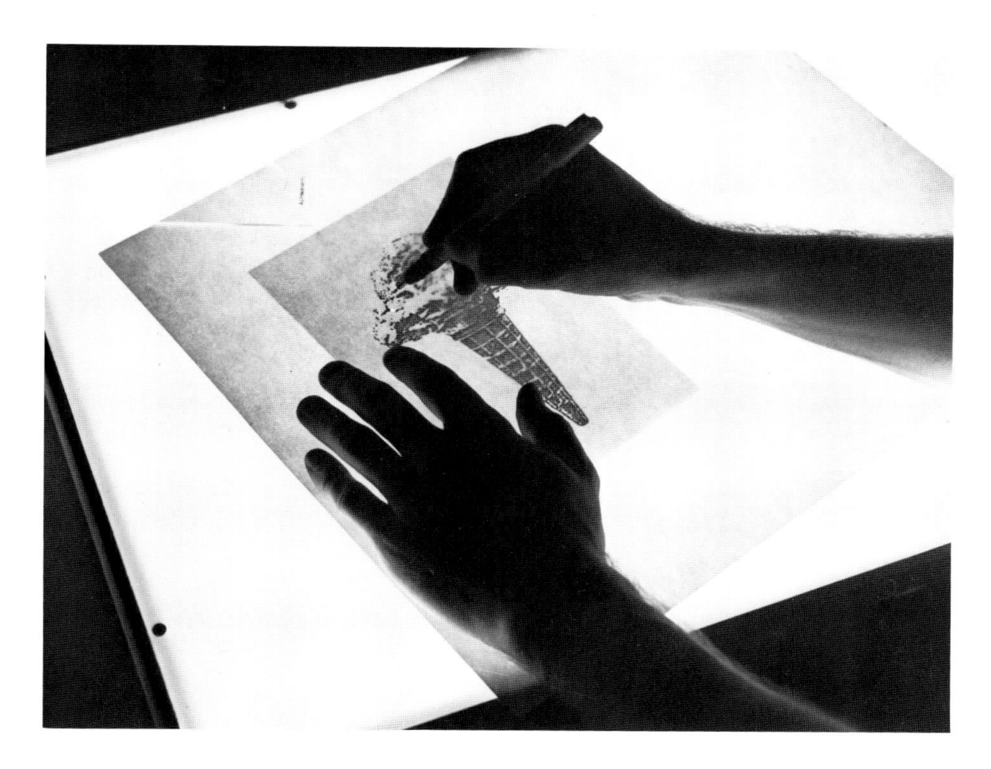

Figure 4.24. Transferring a design by the light-box tracing method.

Visual Estimation Techniques

Visual estimation is the process a designer uses to transfer type or images from one layout to the other by freehand drawing. It is often time-consuming and not extremely accurate but it can be used in cases where it is not critical for a design to be evaluated.

Mechanical Enlarging

Mechanical enlarging is a process that involves the combination of a machine that enlarges or reduces an actual drawing by hand. This method is faster and more accurate than visual estimation enlargement but it does require time for rendering and its accuracy depends on the skill of the designer.

Photographic Enlarging

Photographic enlarging is the fastest and most accurate way to enlarge but can also cost the most, depending on the equipment you use. The most economical method is to use an enlarging and reducing photocopying machine, available in most design firms, schools, and copying shops. The most costly, but the most precise, method is to make a photostat or velox print from your original.

SCALING ARTWORK

Generally the actual size of artwork will be different from the size you need for your layout. Therefore, you will have to *scale* artwork, that is, enlarge or reduce it to fit your design without altering its proportions. There are several techniques you can use to scale artwork.

Diagonal Line Scaling

To scale artwork using the *diagonal line scaling* technique, follow these steps (Fig. 4.25):

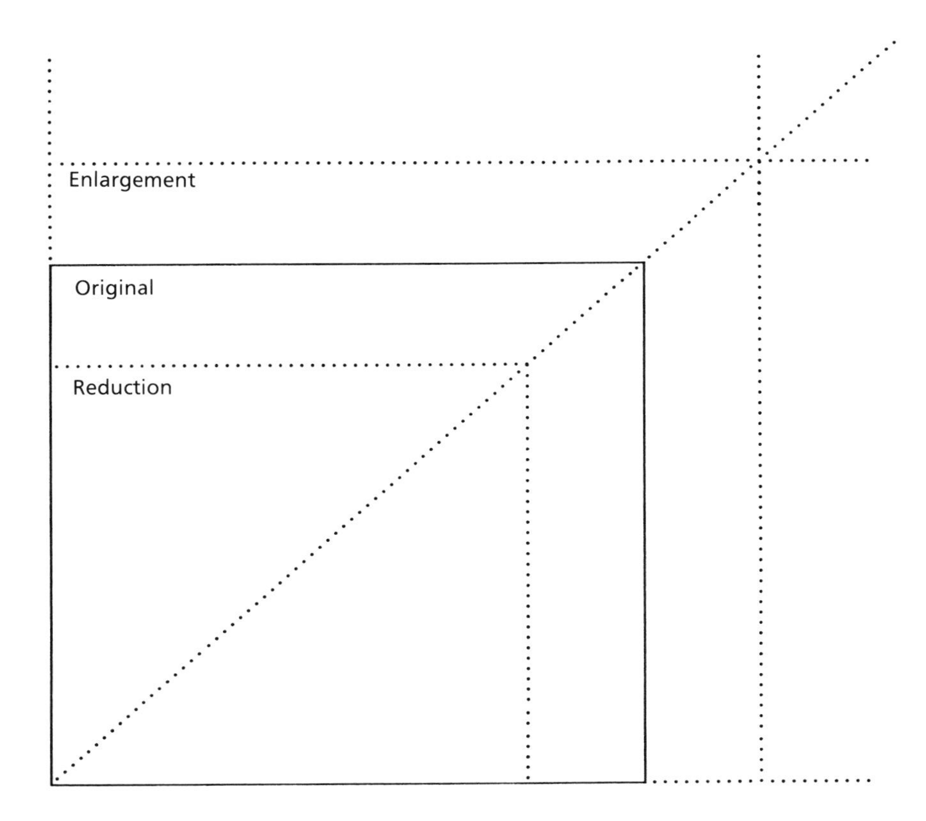

Figure 4.25. Diagonal line method of proportionally enlarging or reducing artwork.

1. Square up the original artwork on the drawing table and tape it down securely.
2. On a sheet of tracing paper taped over the art, draw a rectangle around the area to be enlarged or reduced.
3. Within the rectangle, draw a diagonal line from the lower left-hand corner to the upper right-hand corner.
4. Once drawn, this diagonal line can be extended outside the original rectangle to create a proportionately larger rectangle. To do this, draw a horizontal and a vertical line that meet at the desired point along the extended diagonal line. To make a rectangle smaller than the original, make the point of intersection anywhere along the diagonal line that is inside the original rectangle. Using this method, you can determine how much horizontal and vertical space your artwork will occupy in your layout, and make adjustments accordingly.

Using a Circular Proportional Scale

The circular proportional scale (Fig. 4.26) described in Chapter 3 under Tools is the fastest and simplest way to reduce or enlarge artwork in proportion to its original size. To use the scale, follow these simple steps:

1. Locate the size of the original artwork on the smaller, inside disk of the scale.
2. Locate the size you want the artwork to be on the larger, outside disk.
3. Rotate the disks until the two sizes line up.
4. Read the scale. Notice that a window has been cut out of the inside disk. An arrow pointing inside the window will show you the percentage the art will need to be reduced or enlarged to fit the desired size. The window will also show you how many times the artwork will be reduced.

For example, suppose a piece of artwork is 8 inches wide by 10 inches long and you want to reduce it so it is 4 inches wide. Find 8 inches on the smaller, inside disk and 4 inches on the larger, outside disk. Rotate the inside disk until the numbers align. To find the percentage reduction required, read where the arrow is pointing inside the window; you will see that it is 50 percent.

Recall that the original art is 10 inches long. If you look at the number on the large disk that aligns with 10 on the small disk, you will see what the reduced length will be. In this case, it is 5 inches.

Figure 4.26. Circular proportional scale.

JOINING MATERIALS

When you prepare background areas on comprehensives, you will often need to use several materials next to one another. When you put these materials together, you want a clean, accurate *registration,* or alignment, without distracting seams or layers. The techniques of joining materials are *butting* and *insetting.*

Butting

Butting is the joining together of two pieces of material, such as colored paper or film, closely along one common edge. The steps in butting two materials are as follows (also see Fig. 4.27):

1. Apply rubber cement or spray adhesive to the backs of the pieces of material to be joined. Both pieces should be slightly larger than the desired finished, trimmed size.
2. Place one piece of material over the other so that each overlaps beyond the intended joint and press firmly in place.
3. Using a T square and a steel triangle as straightedge guides, make a common cut through both pieces of material at the desired butting edge with an art knife. Squirt a small amount of rubber cement thinner under the newly cut edge of the top overlapping material to facilitate the easy removal of the excess material. Repeat this step to remove the excess underlying material.
4. Place a sheet of tracing paper over the joint and burnish it firmly to maintain a strong adhesion of the materials along the seam.
5. Moisten a cotton swab or tissue with rubber cement thinner and use it to remove excess rubber cement.

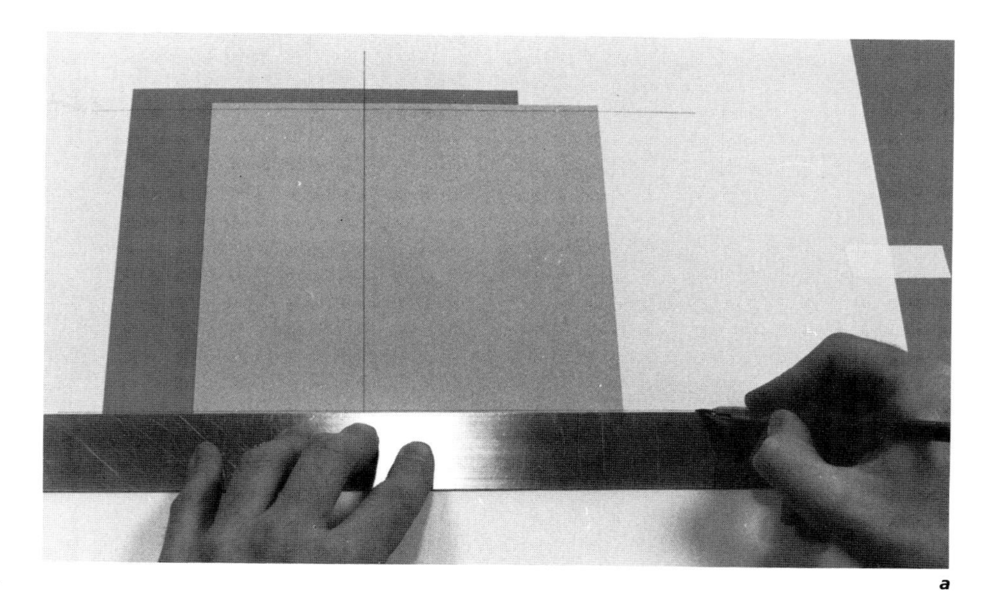

a

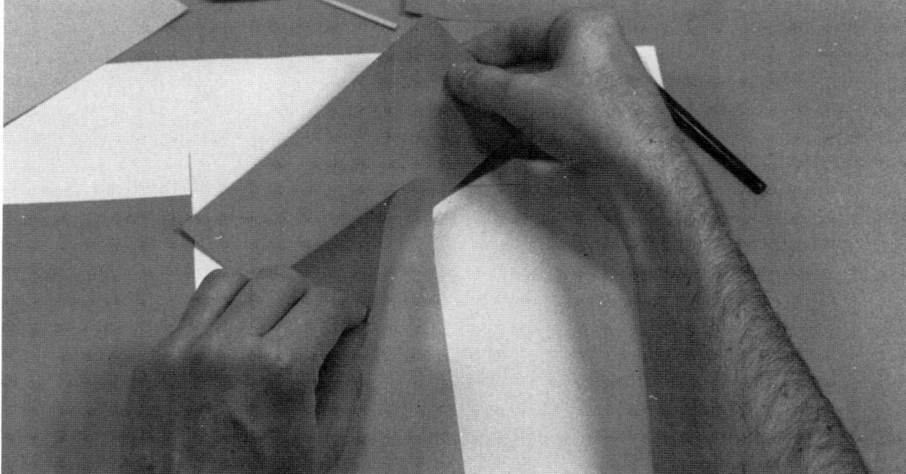

b

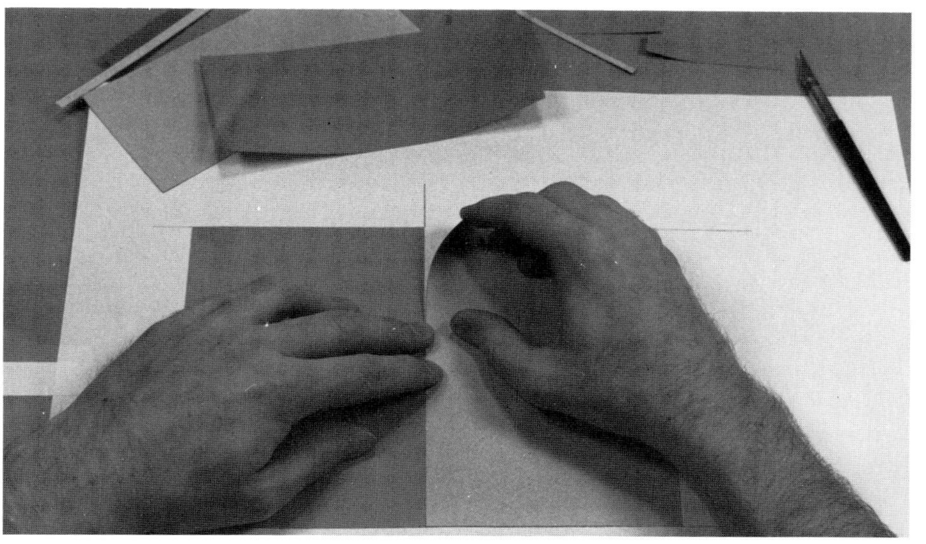

c

Figure 4.27. Butting. *a.* Adhere the paper that is to be "butted" onto a backing board (such as 3-ply bristol board) so that the sheets overlap one another. Next, using a sharp, hard pencil, indicate the trimming size of the piece. *b.* Using a metal straightedge and an art knife, make cuts completely through both layers of paper along the pencil guidelines. Remove all excess paper, including the overlapping piece underneath. *c.* After removing the overlap, burnish down the seam and erase the pencil guidelines.

Insetting

Insetting is the securing of one piece of material, such as colored paper or film, partly or wholly inside another along two or more common edges. Follow these steps to inset a piece of material inside another material (also refer to Fig. 4.28):

1. On a sheet of tracing paper, draw the position of both pieces of material to be joined—the larger piece and the smaller piece to be inset.
2. Hinge the tracing paper over a baseboard, such as illustration board or whatever surface is to be used as a base in making the finished comprehensive.
3. Using rubber cement, adhere the larger sheet onto the baseboard and then adhere the smaller sheet (which should be cut slightly larger than the desired trim size) on top in the position shown on the tracing paper. Work underneath the tracing paper when positioning materials to ensure that they are in the right place before they are adhered.
4. With a pushpin, poke a hole through the tracing paper and onto the top sheet of material. If it is a square or rectangle, just poke a hole in each corner. If it is an irregular shape, poke enough holes so that you have an accurate cutting guide.

a

b

c

Figure 4.28. Insetting. *a.* Adhere the paper that is to be "insetted" onto the base sheet (which has been adhered to a backing board). Next, in pencil, draw the trim size of the intended shape and then cut completely through both sheets of paper. Remove the excess paper surrounding the shape. *b.* Pick up the top (raised) shape and then place it to one side. Next, remove the bottom shape so that the underlying backing board is exposed. Place the top sheet back into its former position. *c.* Burnish the seams thoroughly.

5. Using the holes as a guide, make a common cut through both materials with an art knife. With a cotton ball or tissue, dab a small amount of rubber cement thinner around the cut edge and remove the excess material outside of and underneath the cut shape.
6. Burnish down the seams of the joint and remove excess rubber cement with rubber cement thinner.

PREPARING INTERIM MECHANICALS

An *interim mechanical,* a term invented to differentiate an inexpensive, photomechanically reproduced (as by photostat or photolith) layout of the entire design and all its elements from a final mechanical that is sent to the printer, is used to help prepare the advanced comprehensive (see Fig. 4.29). It contains typeset or reproduction-quality type and finished artwork. Generally, the art is rendered on a separate piece of paper or film, then photostatted or copied, and finally adhered to the interim mechanical board.

Interim mechanicals should be executed on a smooth-surfaced illustration board such as a Bainbridge #172 or a bristol or coated box board. Make sure the board is large enough so that the artwork and other design elements can fit within the crop marks and any necessary instructions can be written outside of the crop marks.

There are four steps that should always be followed when you prepare art for an interim mechanical:

1. Gather the materials and tools you will need. Square up the board onto a drawing table and tape it down with white tape.
2. Using the method described earlier under Scaling Artwork, scale artwork, type, rules, symbols, logos, photographs, and so on, to the sizes they will be on the advanced comprehensive and photostat them at their final sizes.

   Draw all horizontal and vertical guidelines for all the dimensions of the design elements on the board with a nonrepro pen or pencil and rule crop marks for the entire mechanical with black ink using a fine-nib technical pen. The guidelines should be approximately one half inch in length and should be drawn at a distance equal to approximately one eighth inch outside the crop marks.

   Trim the photostatted pieces and adhere them to the board with one-coat rubber cement or spray adhesive. Follow your finished layout for positioning the design elements.
3. After you have adhered all the elements to the interim mechanical board, photostat or photocopy it. Trim the stat or copy to the size indicated by the crop marks. Use this copy to make sure all the elements are in their proper places and to see if any changes need to be made.
4. After the trimmed stat or copy has been evaluated, changed if necessary, and approved, make a photostat or photolith film of the board, which will then be used for the advanced comprehensive.

Figure 4.29. Example of an interim mechanical board being ruled with guidelines and crop marks before type and/or artwork is added.

Preparing Interim Mechanicals with Several Colors

If the advanced comp will have more than one color, you will need to prepare the interim mechnical to accommodate each of the colors. In addition to the steps for interim mechanicals, follow these steps for multicolored mechanicals:

1. Place the line and halftone copy for each color on a separate *overlay,* that is put one color (usually black) on the baseboard and each additional color on a

piece of clear acetate or transparent material on top.

2. Affix self-adhesive register marks on the baseboard and acetate overlays for accurate positioning and assembling of the comprehensive.

3. Make a photostat, photolith film negative or positive of the baseboard and of each separate color. A photolith film negative is a photographic reproduction on a clear film in which the black and white of the original image are reversed. For example, if the original contains black type on a white background, the photolith negative film would render it as a black background with clear type. A photolith film positive is a photographic reproduction where the black and white on the original is duplicated on film.

Figure 4.30. Text type that matches the size and leading of the desired typeface is often clipped from a printed annual report or magazine for use on an interim mechanical as a cost-saving measure. The original is shown at left, the altered version at right.

ALTERING TYPE

When producing photographically rendered comprehensives you may have to alter display or text type to fit your layout. Many typographical effects can be easily and inexpensively achieved in the graphic design studio. Basically, you cut up existing type (as shown in Fig. 4.30), paste it on a board as you wish it to appear, and photostat it to create the look you want. The following are some specific techniques you can use to alter the look of type.

Altering Line Spacing

In some layouts you may wish to change the *line spacing,* or *leading,* between each line of type. Figure 4.31a shows how to remove space between lines, and Figure 4.31b shows how to add space.

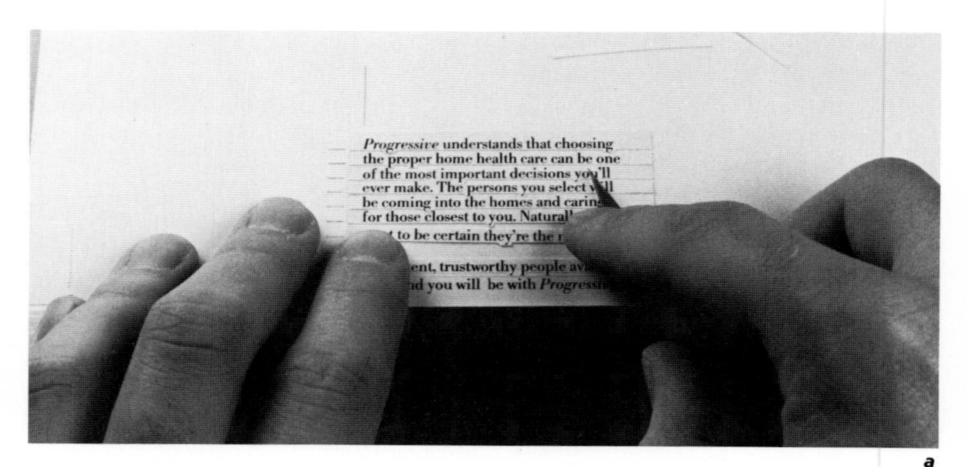

a

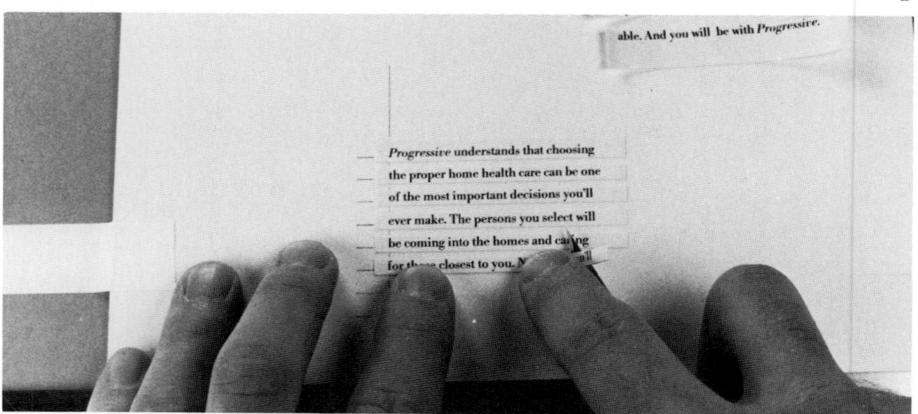

b

Figure 4.31. Altering line spacing: *a.* closing up line spacing; *b.* adding space between lines.

1. Use a type galley, which is a reproduction quality on black and white paper or film output from a typesetting system that contains headline and/or text type, or a photostat of the original type, and apply rubber cement or spray adhesive to the back. Adhere the type to a scrap piece of illustration board. If the altered type will be put directly onto a mechanical, adhere the type to heavyweight bond paper, lightweight bristol board, or multipurpose drafting vellum.
2. Using a T square, square up the mounted type along its baseline, and tape the mounting board onto the drawing table or board. Using a triangle as a straightedge, indicate the left- and right-hand margins in pencil. These margins will be used later to align type vertically after it has been cut apart.
3. If you need to close up the leading between lines of type, line up a steel T square under the first line of type, and with an art knife make two long horizontal cuts that create a removable section equal to or greater than the desired reduction in space. Then make vertical cuts so that the pieces can be removed and discarded. Raise the second line of type so it is the desired distance from the first line. Repeat this step for all the lines of type.
4. To open up the leading between lines of type, make one horizontal cut between each line of type using a steel T square and an art knife. Then, in pencil, mark the line spacing you want: Starting from the baseline of the first line of type and working downward, mark the new, wider spacing on the left side of the type column. Open up the spaces between all lines of type, making sure that each baseline aligns horizontally with the pencil marks and vertically with the column and/or margin.
5. After you have adjusted the spacing, lay a Haberule, or type gauge, over the altered body type by aligning the marks for the leading you want with all the lines of type in the text block and make any final adjustments in spacing. When the spacing is correct, make a photostat of the type and place it on the actual mechanical.

Altering Letter Spacing

Another way to change the look of type is to increase or decrease the space between letters, that is, to alter the *letter spacing* (see Fig. 4.32). To do this,

1. Apply rubber cement or spray adhesive to the backside of the type galley or photostat that is to be altered. Adhere the galley or photostat to illustration board, heavyweight bond paper, lightweight bristol board, or multipurpose drafting vellum, depending on how the respaced type will be used. For example, if you will be pasting the altered type on the interim mechanical, use bristol board; if you will be photostatting the respaced type, mount it on a scrap of illustration board or any smooth white surface.
2. Using a T square, square up the mounted type by aligning its baseline, and tape the board or whatever material you are

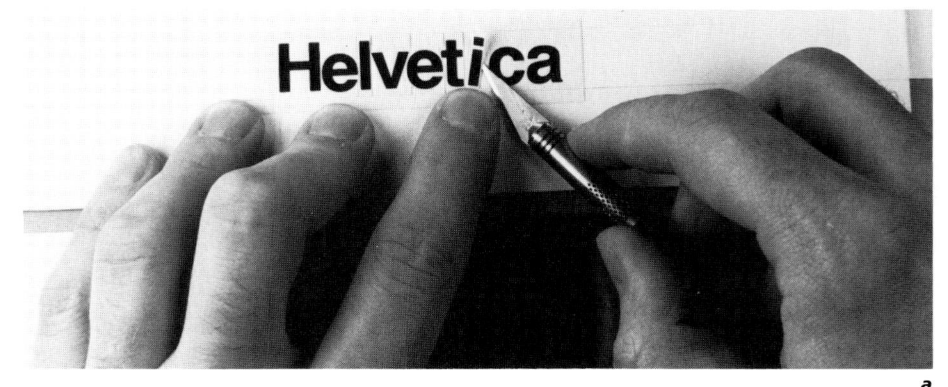

a

b

Figure 4.32. Altering letterspacing: *a.* closing up space between letters; *b.* adding space between letters.

using onto the top of a drawing table.

3. Create a "track" on which you can move the type to the left or right, but not vertically, by making a horizontal cut completely through the photostat or galley above or below the lettering, allowing enough space so that the ascenders and descenders are not cut off in the process. Use a steel T square and art knife to ensure straight, clean cuts. If the type is to be opened up, make sure the cuts extend far enough beyond the type area to accommodate the wider spacing.

4. If you want to close up or tighten the letter spacing, using an art knife and steel triangle to ensure clean right angles to the baseline of the type, make two vertical cuts between each letter to be respaced. If the existing letter spacing is too tight for cuts to be made with a steel triangle and a T square, make the cuts freehand. After cutting, remove extraneous pieces and slide the letters together to the desired tightness.

5. If you want to open up the letter spacing, make a single vertical cut between each letter using an art knife, steel triangle, and a T square, or make the cut freehand. Open the letter spacing to the desired width.

6. After you have adjusted the letter spacing, lay a sheet of heavyweight vellum over the respaced type so that you can evaluate it without looking at distracting cut edges. Make any alterations necessary; then place the type directly on a mechanical or photostat it, depending on the kind of board you are using. For example, if the altered type was mounted on a scrap of illustration board, it should be photostatted first, then placed on the mechanical.

Curving Type

In addition to changing the spacing of type, you can change the shape of its layout (Fig. 4.33). Follow these steps to curve type.

1. Draw the curve you want in pencil on a sheet of tracing paper.

2. Lay the tracing paper over the type galley or photostat of the original type and trace the letters onto the curve either by aligning the top of the letter with the inside of the curve or by placing its baseline on the outside of the curve and turning the paper continually to align the type.

3. Remove the tracing paper. Apply rubber cement or spray adhesive to the back of the galley or stat. Using an art knife and steel T square, make a horizontal cut above and below the letters on the type galley or photostat.

4. Make partial slits with an art knife either above or below each letter so that the type will be easier to bend around a curve. If the type will follow a circular shape that faces upward, have the slits extend downward from the top edge of the galley or stat to about the baseline of the type. Additionally, make a series of "V"-shaped cuts below the baseline. If you want the shape to face downward, make "V"-shaped

Figure 4.33. Slits are made in the type sample to facilitate bending it around an upward curve. When type has to be bent around a very severe curve, letters may have to be cut out individually. To more easily position lettering around a curve, use a tissue layout as a guide. Straight type is often curved so it can fit into a smaller space.

cuts so that they extend upward from slightly above the baseline of the type to the top edge of the galley or stat, and make slits downward from slightly below the baseline of the type. If the circular shape is very tight and the type cannot be bent to match it, you will have to cut the letters apart and position them separately.

5. After the type galley or photostat has been cut, position the lettering under the tissue layout in order to follow the curve. Use a pair of tweezers to help grip the edges of the type during positioning. Burnish the type onto the mounting board.
6. Remove the tissue layout and place a sheet of heavyweight vellum over the repositioned type so that you can evaluate it without seeing distracting cut edges. After making any adjustments needed, place the new type on a mechanical or photostat it, depending on what it is mounted on. For example, if the type was mounted on a sheet of lightweight bristol board or multipurpose drafting vellum, it can be placed directly on the mechanical.

Figure 4.34. Matting is a process in which an opening, or "window, is cut out of a mat board and then dropped over a comprehensive to form a border or frame. Mats are most often used to conceal any stray marks on underlying art work, to support an otherwise flimsy piece, or to hold films in place over rendered backgrounds.

PROTECTING AND ENHANCING COMPREHENSIVES

Professionally rendered two-dimensional comprehensives are almost always presented either matted or mounted. Both of these methods protect and enhance the comp, but each is prepared slightly differently.

Briefly, *matting* is a process of opening up a window in a mat board, which is then dropped over the finished piece to form a border (Fig. 4.34). Mats are used to conceal stray marks or to support flimsy, hard-to-handle pieces. Mats are most commonly used to sandwich a film

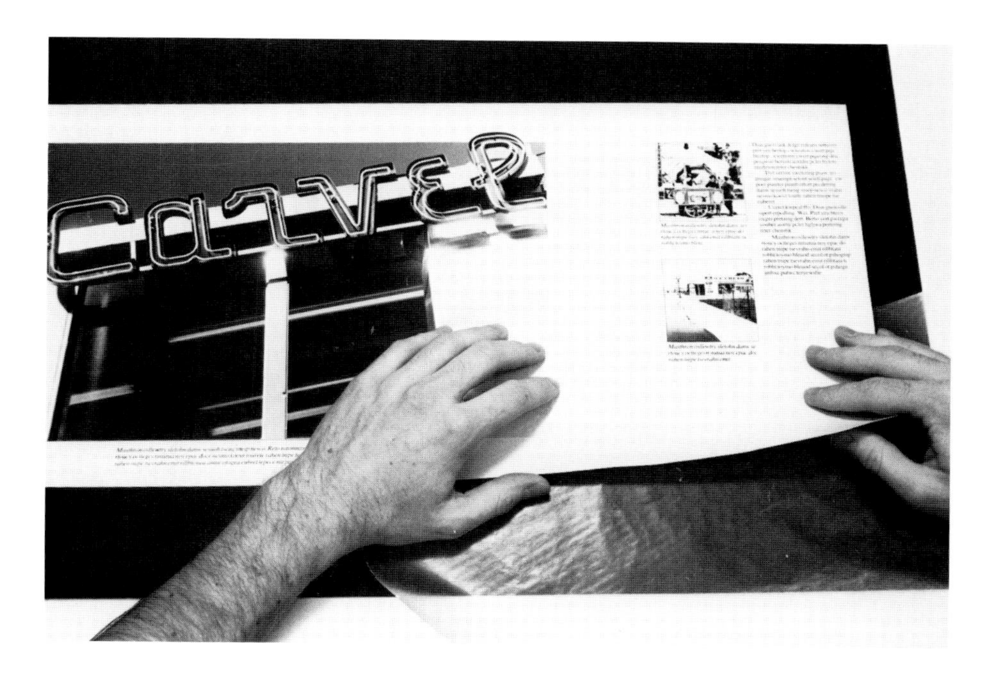

Figure 4.35. Mounting is a process in which a two-dimensional piece is trimmed and adhered onto the surface of a mat board.

overlay of type and artwork over a rendered background.

Mounting is a process of trimming a two-dimensional piece and adhering it to the surface of a mat board with rubber cement or spray adhesive (Fig. 4.35).

The choice of whether to mat or mount a piece is usually determined by the medium on which the artwork was originally rendered (see Fig. 4.36). For example, if you rendered your comprehensive on illustration board and it has many stray marks, a mat is the best way to make a clean presentation. If you rendered it on thin, opaque paper or photostat paper, it should be trimmed flush to the outside edge of the artwork and mounted on a mat board. If you rendered your design on flimsy drawing, layout, or visualizing paper, you could either mat it by sandwiching the artwork between opaque white board and a mat board from which an opening has been cut, or back it with a thin, opaque material such as heavyweight bond paper and then mount it on a mat board.

With either matting or mounting, a border is created around your artwork. Under no circumstances should the mat or mounting board distract attention from the art. Mounting is often preferable since that process does not add another board to the piece (see Fig. 4.37).

Figure 4.36. The medium on which the artwork was originally rendered will usually determine whether matting or mounting is most suitable. For example, a window mat is used here since type has been rendered on an acetate overlay.

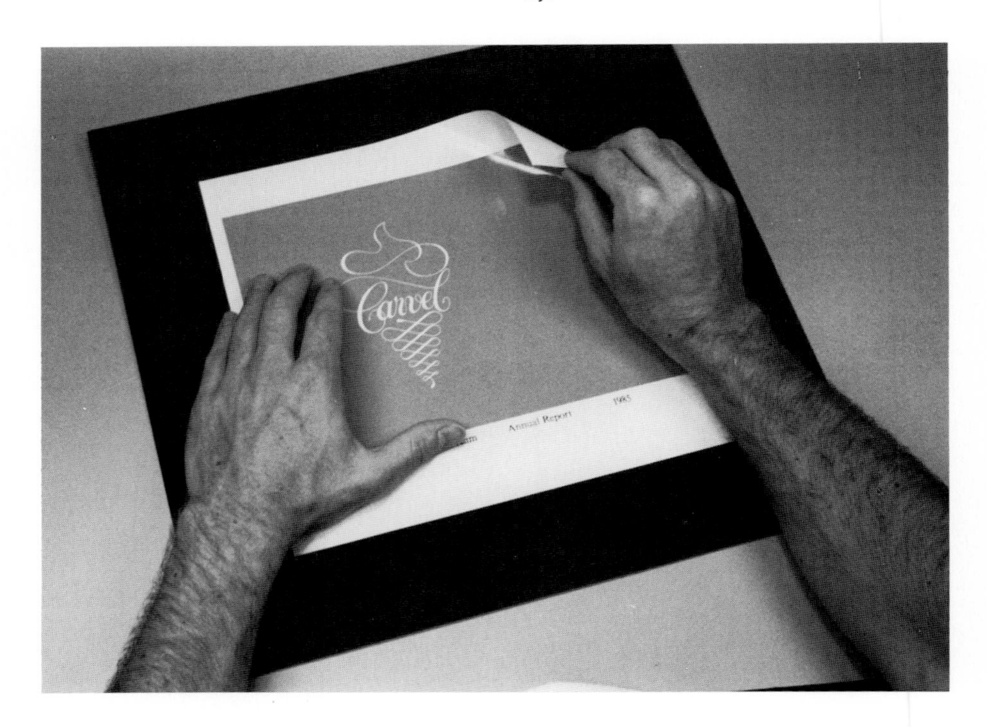

Figure 4.37. For most general applications, mounting is the cleanest, most suitable method, since it does not add on another board, which can be distracting, to the piece.

Matting

Here are the steps to mat a piece of artwork,

1. Using a steel T square and triangle as straightedge guides, square up the artwork and tape it securely to the top of a drafting table.
2. Place a sheet of tracing paper over the artwork and tape it to the tabletop or outside edges of the artwork. On the tracing paper, mark off the dimensions of the mat opening in pencil.

The mat opening should be cut exactly to the size of the art. A common mistake is to leave a border around the art, making the border look like a part of the design.

3. Place the tracing paper over the mat board. Using a pushpin, puncture holes into the mat board to mark the four corner points of the mat opening. Remove the tracing paper and connect the points with a light rule using a sharpened 2B pencil. This rectangle indicates the mat's inside opening.
4. To establish the mat's outside dimensions, measure out 2 or 3 inches from the rectangle and lightly connect those points. You now have a rectangle within a rectangle.
5. Using a sharp utility knife and steel straightedge, cut out the mat opening (the inside rectangle) following the pencil guidelines. To achieve clean, sharp corners, begin each cut at one corner pinhole and extend it to a point just short of the next pinhole. Do this for all four cuts. Then rotate the mat board and repeat the cuts in the opposite direction, completing the unfinished cuts.
6. Following the pencil guidelines, cut the outside dimensions of the mat with a mat knife.
7. Using a sanding block, lightly sand all the cut edges to remove rough or uneven spots.
8. Curl two 1-inch pieces of white tape with the sticky side exposed to form closed "O" shapes and adhere them to the back of the mat. Place the mat over the artwork in the desired position and press firmly so that the tape sticks to the artwork.

   If the artwork was rendered on board that is larger than the outside dimensions of the mat, the board must be cut smaller. Trim the board so that it is slightly smaller than the mat's outside dimensions but larger than its inside opening. This will allow you to conceal the board containing the artwork and will allow you to secure the artwork to the back of the mat board with tape later on.
9. Carefully turn over the mat board (which is attached by tape to the artwork) without shifting the position of the mat's opening. Both the mat's opening and the artwork are now facing down. Tape the artwork securely onto the mat board on one side, forming a hinge. Lift the artwork and remove the pieces of curled white tape that you attached earlier. Tape the remaining sides of the artwork securely to the mat board.

Mounting

The following are the steps to follow to mount a piece of artwork.

1. Using a steel T-square and triangle as straightedge guides, square up the artwork and tape it securely to the top of a drafting table.
2. With an art knife, trim the artwork to the desired size.
3. Turn the artwork over and apply a coat of rubber cement to the back and set it aside to dry.
4. On a piece of mat board 2 or 3 inches larger than the artwork on all sides, outline the dimensions of the trimmed artwork in pencil.
5. Apply rubber cement to the mat board in the area outlined in pencil and let it dry.
6. Place the artwork cement side down onto a nonstick wax paper slip sheet so that it extends about one half inch beyond the top edge. Position the artwork on the mat board in the area outlined in pencil and tack the adhesive-backed artwork along its top edge. Pull out the slip sheet and press the artwork firmly onto the mat board. Next, place a sheet of heavyweight vellum over the entire piece and burnish it with the side of a plastic triangle to remove bubbles and ensure that the artwork lies flat.

   The slip-sheet method of placing adhesive-backed type or artwork onto another surface allows you to precisely position these elements before adhering any element permanently.
7. Square up the mat board so that the artwork aligns along a T square and tape it securely to the top of a drafting table. Lightly draw a 2- or 3-inch border around the artwork with a sharpened 2H pencil.
8. Trim the mat board along the border guidelines with a mat knife; discard the excess board.

PREPARING DUMMIES

When you are working on a design that will be produced in a format other than a flat surface, it is necessary to prepare a *dummy,* a three-dimensional comprehensive that is folded into shape to show all parts of the design. Actually, a blank dummy is prepared first. A blank dummy is any three-dimensional form that already exists or has been constructed out of paper or board or another material and is used as a base on which graphics (your layout or comprehensive) are wrapped or applied (Fig. 4.38). It can be a container that has had its label removed or covered

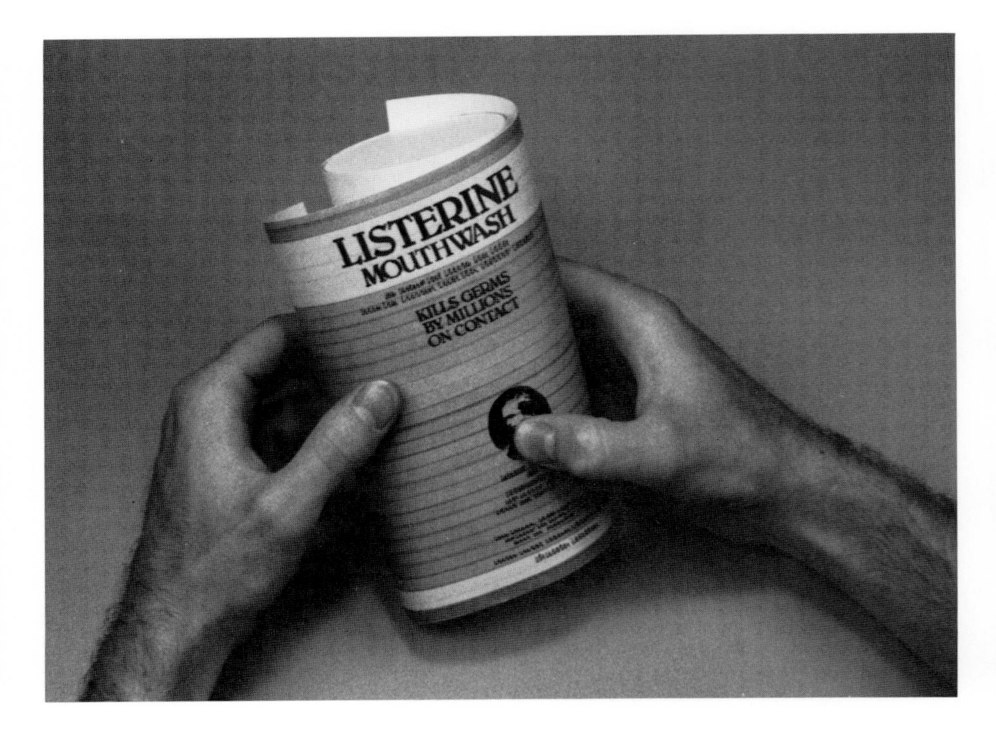

Figure 4.38. Blank dummies are used to wrap flat sketches or artwork in order to make an accurate assessment of their effectiveness as three-dimensional designs.

over or it could be made to simulate the materials that would be used in printing, such as wood; glass, paper, plastic, or cardboard. A blank dummy is used for mounting finished layouts, comprehensives, and advanced comprehensives. The easiest way to make a blank dummy is to take apart an existing printed piece, trace its shape lightly in pencil, and cut out the traced shape with an art knife. Dummies are made for items such as packages, folders, and books.

There are two basic techniques needed to render almost any dummy —scoring and folding.

Scoring

Scoring is a process of indenting heavy paper stock or board, allowing you to make a clean, accurate fold. As a general rule, the thicker the material, the wider the score you will need to make. Paper clips and burnishers are good for scoring. Large paper clips provide a wide score and small paper clips make narrow scores on thin materials. In some cases, where the material is very thick, a shallow cut can be made first with an art knife to help facilitate folding.

When you work with paper, always make sure the surface you scored ends up on the outside of the folder or booklet for minimum paper stretch (Fig. 4.39). When you work with thick, heavy boards, begin by partially cutting the surface using an art knife and straightedge. Then fold as you would fold thinner paper. If the board still will not fold flat, make the cut a little deeper. When scoring by hand with a burnisher or paper clip, always use a straightedge to ensure a clean fold.

Folding

In the production of dummies, *folding* is a technique by which a single piece of printed paper is folded one or more times. There are two basic folds (shown in Fig. 4.40): The first is a *parallel fold,* which is a fold made parallel to a previous fold. This method is used for folding business letters to fit within a standard-sized envelope. If you alternate the direc-

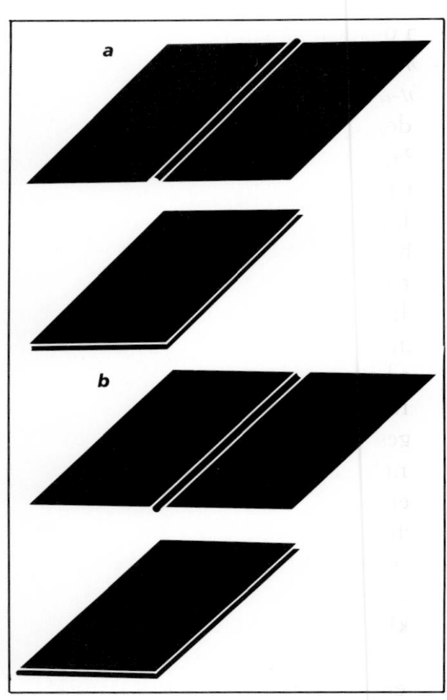

Figure 4.39. Correct *(a),* and, incorrect *(b)* folder-scoring procedures. Folders should always be scored so that when they are folded the ridge or hinge is on the inside, allowing minimum paper stretch.

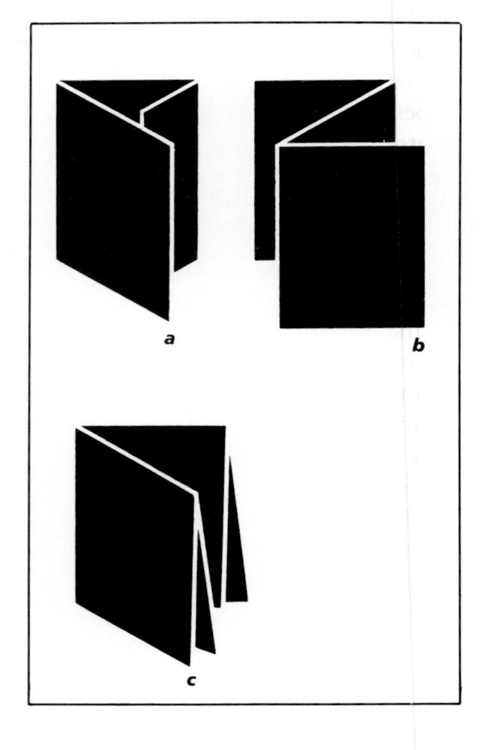

Figure 4.40. Assorted folder styles: *(a)* parallel, *(b)* accordion, *(c)* right angle.

tion of parallel folds you get an *accordion fold.* The second fold is a *right-angle fold,* which is made at a 90-degree angle to a previous fold.

Paper *grain,* which is the direction that paper fibers align themselves during paper-making, affects the ease with which you can fold paper and boards. A scored fold is easier to work with if it is made across the grain. This is because when paper or board is scored across the grain, the fibers are transformed into multiple hinges that allow easy folding while maintaining paper strength. When paper is scored parallel to its grain, on the other hand, the fibers separate and weaken.

Making Package Dummies

When you are developing a design idea that will be executed as a three-dimensional package, such as a box of cereal or a can of fruit, you need to make a package dummy so that the idea can be evaluated. It is difficult to judge how a piece will look unless it is actually wrapped onto the package. Putting the layout into its final shape is the best way to see the juxtaposition of panels containing different shapes, colors, and typographical elements. In fact, in most cases the shape of the package will dictate the best arrangement of the design elements.

When rendering graphics for a package dummy, render them flat on the material that will be used for production, or on a similar material, and then transfer the design to the dummy. For example, if the design is to be reproduced on an aluminum can, you can render the dummy on dull foil, trim it to size, wrap it around a can or hand-crafted cylinder, and fasten it into place with spray adhesive or double-sided tape. If the package is a rectangular folding box (see pages 85–86), you can trim the layout, adhere it to the flat box, and score, fold, and assemble it later. Or you can cut the layout into sections and adhere the parts to the individual panels of either a folding or set-up box (see pages 85–86). A dummy made by adhering the entire layout is less distracting and more easily evaluated, but one made by adhering parts of the design to panels is easier to change.

In many cases, the client will give you the box, cylinder, or other type of container to use for the dummy. If the container is printed, either cover it with a neutral opaque background material such as a white, adhesive-backed paper (Pantone coated paper is a good choice), spray paint it or, if possible, turn it inside out. If necessary, you can trace the container, then cut, score, and fold it into the desired shape. If the container is plastic, you might be able to remove any printing or graphics with a cotton swab moistened with acetone. Although packages can take almost any form, the most common shapes are rectangular boxes and cylinders.

Boxes

There are basically two types of boxes used for packages: folding boxes and set-up boxes (Fig. 4.41). Make your dummy from the type of box that will be used for the final product.

The folding box, which is the most common, is made of one piece of white folding box board that has been scored and die-cut to form a specific

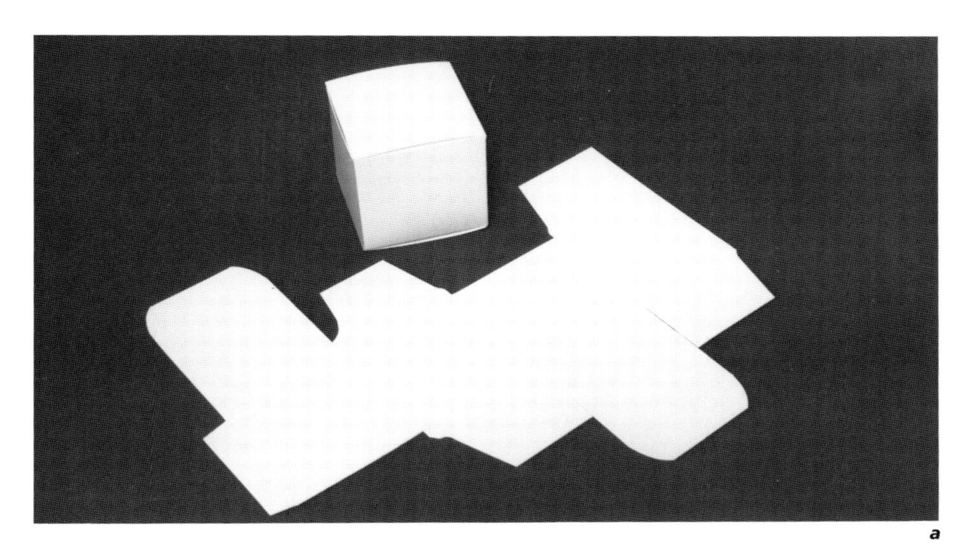

a

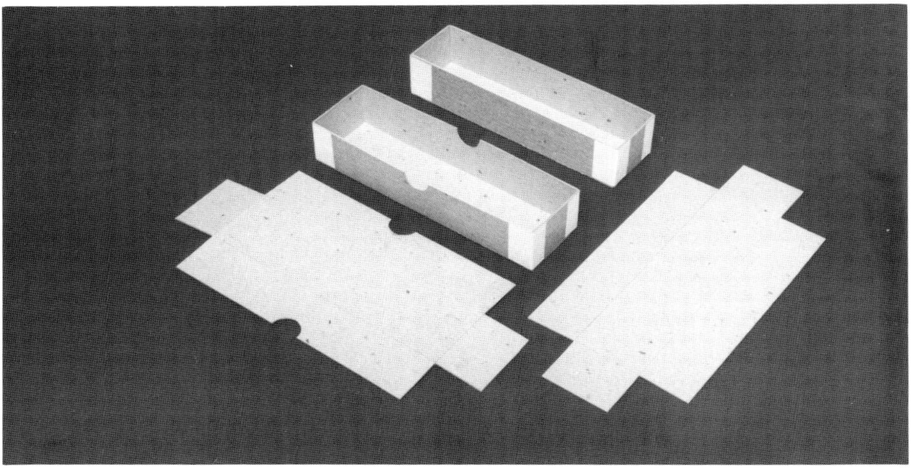

b

Figure 4.41. Blank box dummies. A folding box *(a),* shown assembled and flat, and a set-up box *(b),* shown assembled and flat.

shape when folded. *Die cutting* is the printing procedure by which a shape is cut out of paper or board with special steel rules before a piece is scored or folded. These boxes are designed so they can be printed, cut, scored, and glued in one operation and then shipped flat to the company that manufactures the contents, where the box is then assembled.

The set-up box is usually made of two pieces of rigid gray-colored chipboard, one for the base and one for the lid. It is assembled by first cutting, scoring, and folding the gray chipboard into the desired size, then securing the corners with paper, fabric, or metal reinforcements, and finally wrapping the printed paper around the box. Unlike the folding box, the set-up box is shipped assembled. It is expensive to produce, but it is generally used for expensive or fragile items such as jewelry or chocolates that need a rigid container to prevent movement of the contents (Fig. 4.42).

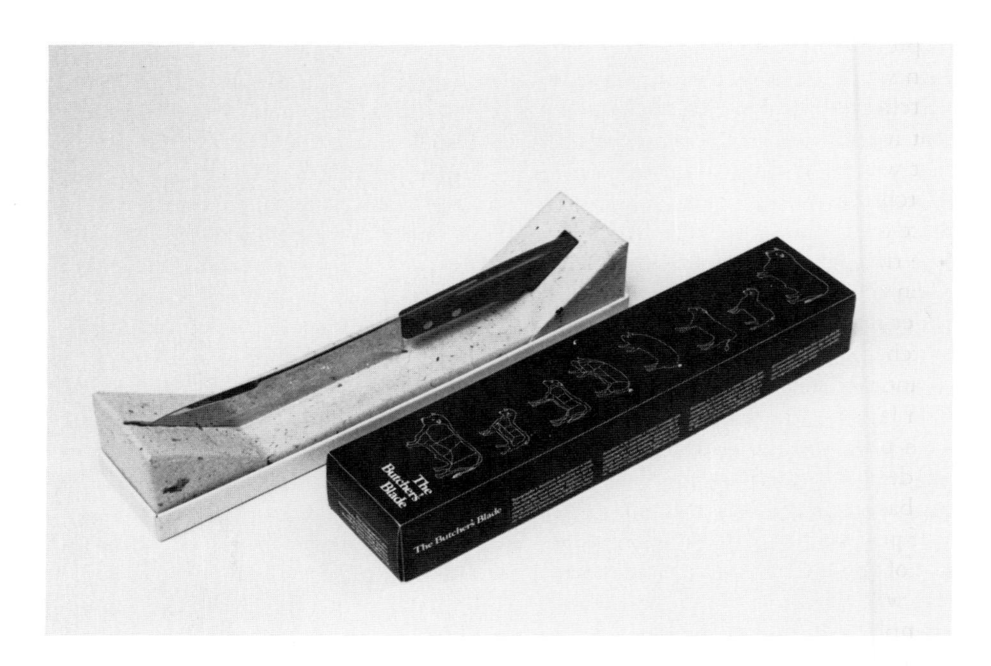

Figure 4.42. A set-up box, whose construction allows the contents to be protected. Note that the base has been wrapped with printing paper and the lid with a photostat.

Cylinders

Cylinders used for packaging come in various sizes, shapes, and materials or can be tailor-made to suit a particular need. Ready-made cylinders include preprinted bottles, jars, and cans; custom-made cylinders can be created from paper, wood, plaster, or any material that will simulate the look of the reproduced piece (Fig. 4.43). You can either apply the type and graphics directly onto the blank cylinder or first apply the design on paper or acetate and then attach it to the cylinder with double-stick tape or adhesive.

Figure 4.43. Ready-made cylinder *(left)* supplied by the manufacturer; homemade cylinder *(right)* produced by the designer.

Making Folder Dummies

A *folder* is essentially a printed piece of paper that has been folded one or more times, making it three-dimensional. Examples include brochures, invitations, greeting cards, and promotional pieces. As with packages, the comprehensive for a folder should be prepared in dummy form. In fact, even when you are doing thumbnail sketches of a folder, keep in mind that it is a three-dimensional piece that will be used "in the round." Sketch the various panels of your piece on one sheet of paper to make sure the elements align properly when the folder is created. If the folder will be printed on two sides, sketch the panels for the second side on another sheet of paper. You can then lay one sheet over the other and push pins through both to make sure the design elements align properly on the back and front sides of the folder.

If possible, render the comprehensive of the folder on the same paper that will be used for printing (for example, see Fig. 4.44). If the paper is not available, use one that is similar in weight, texture, and color so that you can gauge its flexibility and stiffness. If you rendered the design on tissue or layout and visualizing paper, adhere it to the folder paper before preparing the dummy. When all the design elements have been rendered, you can create the dummy: Cement the design to the folder paper (if necessary), cut the folder to size with an art knife, score it with a paper clip, the plastic end of a Zipatone burnisher, or a putty knife, and then fold it into shape.

Care should be taken to ensure that the completed folder dummy can withstand the stresses of folding and handling. You can increase the structural life of your dummy folder by careful cementing, trimming, and scoring. To protect a comprehensive of a folder from getting soiled, apply a generous coat of spray fixative.

Making Booklet Dummies

A *booklet* consists of two or more printed pieces of paper that have been folded and bound together by one of the following methods (see also Fig. 4.45):

Saddle-wire stitching, in which staples (or stitches) are inserted first

Figure 4.44. Brochure featuring a photocopied image on cover stock that has been scored and folded.

Figure 4.45. Popular binding methods (left to right): Velo binding, plastic comb binding, wire ring binding, perfect binding, saddle stitching.

through the spine of the booklet and then through the inside pages to the center fold where they are bent flat.

Side-wire stitching, in which staples are inserted from the front cover of the booklet through all of the inside pages to the back where they are bent flat.

Mechanical binding, in which multiple holes are drilled or punched through the binding side of the booklet cover and all the sheets, and metal or plastic rings or coils are inserted through the holes.

Perfect binding, in which the backs of the collated press sheets (or signatures) are cut off—exposing and roughening the page edges—and adhesive is then applied along the spine and covered with a strip of reinforcing material. A cover is then attached. Perfect binding is used for both paperback and hardcover books.

Booklets are designed in spreads, that is, two pages that face each other. When doing thumbnail sketches of a design for a booklet, show several two-page spreads, being sure to indicate the pagination. To sketch spreads rapidly, begin by drawing boxes, proportional in size to the booklet, to represent a sequence of pages in the booklet. Then sketch your design ideas within the boxes.

Render rough layouts and finished layouts as actual size double-page spreads on tissue or layout and visualizing paper. You can then cement the layouts to a blank dummy made from bond paper. Or you can photocopy the full-size tissue layout, then, using spray adhesive, cement two copies together back to back. If you use a photocopying machine, copy the layout onto a printing stock that is similar in weight, texture, and color to the one used for reproduction. A limitation of this method is that the type and images will be in black. Whatever the method used, create a booklet from the pages by putting them together with saddle-wire stitching or staples.

Booklet comprehensives can be presented as flat two-dimensional spreads mounted onto a mat board or as a three-dimensional dummy in the form of a booklet. Booklet dummies, which enable people to evaluate the visual rhythm and pagination, include text, graphics, and photographs in position, exactly as they will appear when printed (Fig. 4.46).

When preparing booklets, you also have a choice of showing sample pages or the entire booklet. Usually your decision is based upon the type of booklet required and the time and budget available to you. For example, for catalogs in which the page format is generally the same throughout, several spreads are sufficient. For newsletters in which there are a maximum number of pages and an established format, and for annual reports and promotional booklets you would produce a complete booklet to see that all the type and graphic elements fit.

In any case, render the comprehensive flat, using a *grid* to ensure consistency and alignment among the pages. Your grid should be drawn on a sheet of tracing vellum or on clear or translucent frosted acetate and should indicate guidelines for all margins, the column or columns' length and width, position of folios, headlines, or any other item that repeats from page to page, such as a logo or artwork. By placing the grid over each page or spread of the comprehensive, you can check the alignment of the design elements. Render the comprehensive on the paper that will be used for reproduction of your booklet or on one that is similar in color, texture, and weight, using media that closely simulates the printed piece.

If you are presenting a booklet comprehensive as flat page layouts, render them as double-page spreads. For example, if the booklet will be 8½ by 11 inches, the two-page spread will be 11 by 17 inches. If part of the design will *bleed,* that is, go into a margin and off of the page, make sure the design extends outside of the trim area of the spread. After you have trimmed the spread, rubber cement it onto a mat board whose color contrasts sufficiently with that of the booklet.

Clients often request to see a complete dummy of a booklet. The fol-

Figure 4.46. A photocopied booklet dummy using screened photographs and graphic elements that introduce a middle gray value.

lowing procedure is a simple and fast method to produce a 24-page plus cover 8½-by-11-inch blank dummy, which can then be used as a base on which trimmed pages of the finished layouts or comprehensives can be attached. The only materials necessary to produce the blank booklet dummy are 3M Micropore paper surgical tape, 11-by-17-inch white copying paper, and a 70-pound cover stock (or a close equivalent). Refer to Figure 4.47 as you read these steps.

1. Individually fold 13 sheets of 11-by-17-inch white copying paper in half so that you have 26 8½-by-11-inch panels.
2. Lay two of the folded sheets side by side so the 11-inch sides are butted together and the crease in the middle of each sheet is pointed upward. With white tape, secure each sheet onto the drawing board to prevent unnecessary movement.
3. With scissors, cut a 12-inch strip of 1-inch-wide 3M Micropore paper surgical tape. Apply the precut tape to the entire length of the butted

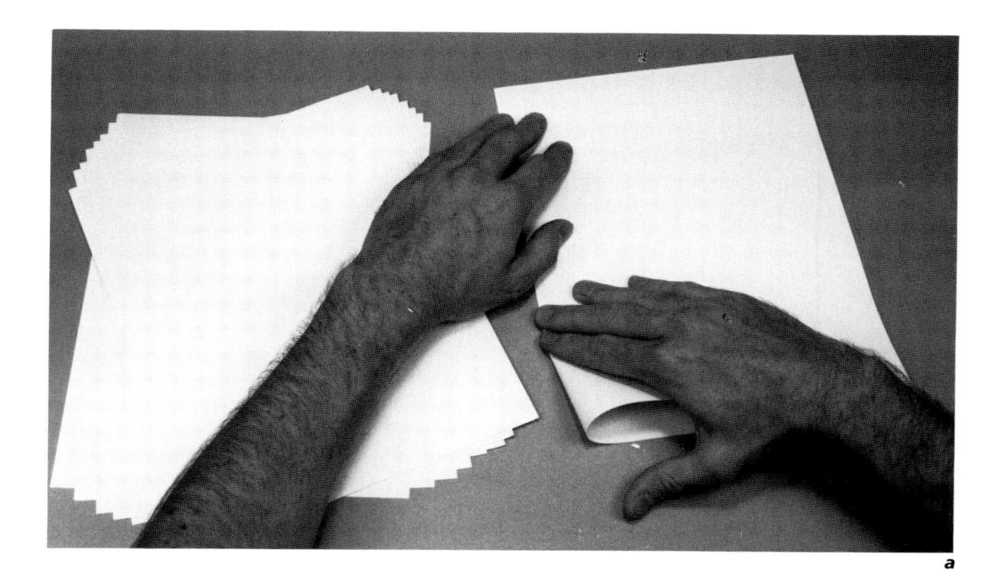

a

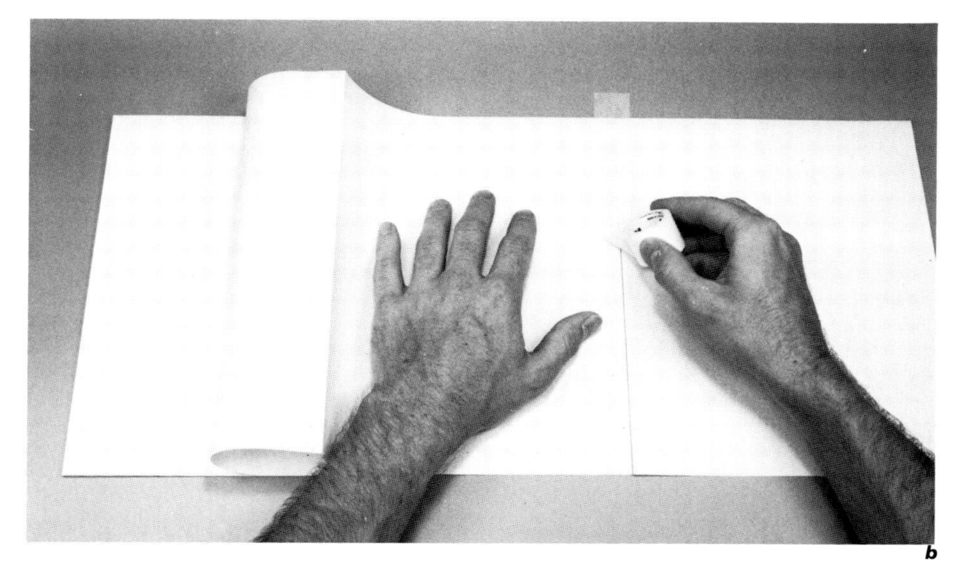

b

c

Figure 4.47. Making a blank booklet dummy. *a.* First, thirteen 11″ × 17″ sheets are individually folded in half to create 26, 8½″ × 11″ panels. *b.* Surgical tape is applied so that it equally straddles the 11″ sides of the butted 11″ × 17″ sheets. The seam is then burnished and the excess tape is trimmed off. *c.* After all the sheets have been linked together in this manner, they are stacked so that the crease of each is facing left and the hidden taped seams are facing right. *d.* After the cover has been scored, it is folded into its proper shape. Double-sided tape is applied to the inside flaps in order to adhere the cover to the inside pages. The finished, assembled blank dummy will resemble a large, continuous accordion. *e.* When executed properly, booklet dummies containing text and illustrative material adequately simulate a printed, bound piece. This gives the viewer the advantage of being able to see the design of each individual spread as well as the relationships between spreads throughout the piece.

d

Figure 4.47 *(Continued)*

e

edge of the two sheets of paper, making sure that the tape straddles both sheets equally. Burnish the seam to ensure strong adhesion; then trim away excess tape with an art knife.

4. Join the rest of the folded sheets following steps 2 and 3. All of the sheets will then be linked, together forming 26 8½-by-11-inch pages. Turn over the joined, folded sheets so that the tape is concealed on the back and the creases point downward. Then stack the sheets so that the creases face left and the hidden taped seams face right.

5. Choose a cover stock to bind, protect, and enhance the booklet. A 70-pound stock is suitable for most purposes.

6. Mark guidelines for cutting on the cover stock. First, mark the 11-inch vertical dimension of the booklet. Then add about 1⅜ inches to the 17-inch length (to accommodate the spine and some overlap on the inside front and back cover) and mark this total 18⅜-inch horizontal dimension.

7. Cut the 11-by-18⅜-inch shape from the cover stock. Prepare the cover stock for scoring by marking the following measurements at the top and bottom edge of the 18¾-inch length in pencil, starting at the left side as follows: ½, $8\frac{9}{16}$, ¼, $8\frac{9}{16}$, ½ inch.

8. Using a paper clip or similar tool, score a vertical line between each of the top and bottom measurements. Fold the cover to the booklet's shape, leaving ½-inch overlaps.

9. Apply a strip of double-sided tape to each ½-inch overlap.

10. Adhere the first spread to the folded overlap (inside front cover) so that the folds of all 11-by-17-inch sheets are facing left. Next, turn the sheets over and attach the back cover to the last spread. To make sure the cover will remain attached when the booklet is open, apply pressure along the edge where the cover attaches to the inside pages.

chapter 5

Advanced Techniques in the Graphic Design Process

In addition to the basic techniques in the graphic design process described in Chapter 4, there are further techniques you can use, which allow you to produce, enhance, and protect advanced comprehensives, simulate printing processes, and produce special effects. The equipment, tools, and materials used are those described in Chapter 3.

PRODUCING ADVANCED COMPREHENSIVES

Advanced comprehensives are generally produced using photocopy, photostat, and other photographic equipment (see Fig. 5.1). With this equipment and advanced techniques, however, you should be able to achieve special effects in the studio that are indistinguishable from those achieved in printing.

On a more practical note, many techniques described here add substantial costs to the final printed job, such as embossing, so it is wise to check with a printer before finished comprehensives are made and presented. Also, double-check that the effects shown can actually be achieved in printing. For example, if the printed piece is a bread wrapper that will be reproduced with Flexographic printing, avoid using delicate type, detailed artwork, subtle gradations or complicated color registration, which cannot be reproduced by this process.

Photocopies

One of the fastest, most convenient, and economical methods of producing (or duplicating) advanced comprehensives is by photocopying them. You can use a photocopy machine to provide clean, opaque images of your design on same-sized sheets for a uniform presentation (Fig. 5.2). Also, the copies can be cut and pasted quickly and inexpensively, making this process an excellent way to provide a finished look.

Photocopying also provides an easy and inexpensive way for you to experiment with your designs (see, e.g., Fig. 5.3). To make color studies, simply draw an outline of the elements in your design on a sheet of paper, make several photocopies of it, and then fill in the outlined shapes on the copies in different colors. To evaluate how all the design elements on your interim mechanical will look on the same surface before applying advanced comprehensive techniques, simply photocopy the mechanical. To see how your layouts will look in three-dimensional form, wrap a photocopy of it around the package, book, folder, or item you are designing for. You can also use the enlargement and reduction capabilities available on some photocopy machines to help you size artwork.

You can use color with photocopies in several ways. After making a black-and-white photocopy of your design, you can add color to it using various materials, such as adhesive-backed colored film, colored pencils, or colored markers. Or you can cut colored paper, such as Pantone coated and uncoated papers, or actual print-

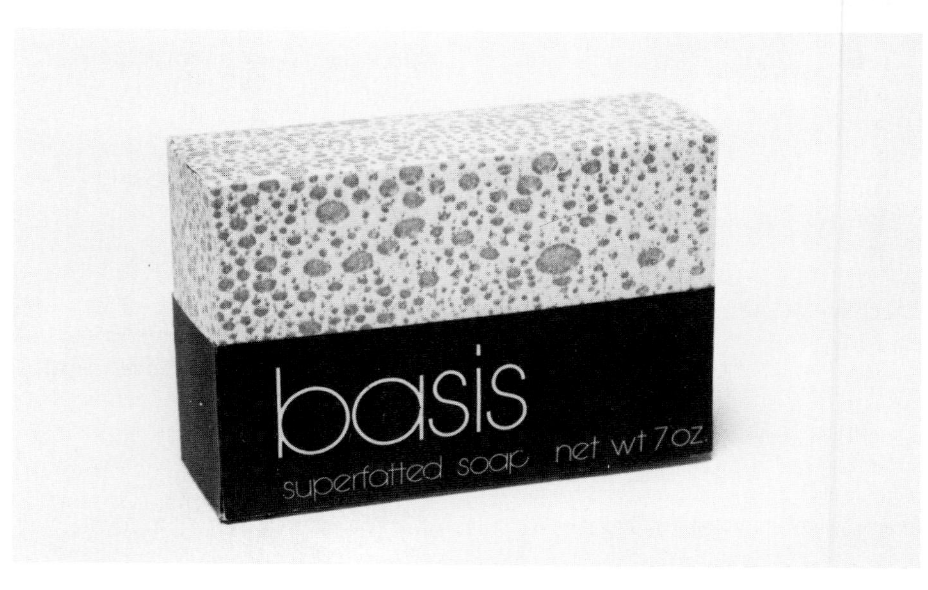

Figure 5.1 *(top).* The equipment used for the production of advanced comprehensives is generally unaffordable for purchase by most independent designers, but there are many suppliers located throughout most urban areas who provide such services (Schaedler/Pinwheel Inc., New York City).

Figure 5.2 *(middle).* The cover and all inside pages of these designs were rendered entirely on a photocopying machine. (Artwork: John M. Obert)

Figure 5.3 *(bottom).* A black-and-white photograph was photocopied on a colored-copying machine using only one color, cyan, for the top panel; the bottom panel was rendered with white dry-transfer lettering (Letraset) on Color-Aid paper. (Artwork: author)

ing papers to the size used in photocopy machines, insert them in the paper feed tray or hand-feed slot, and copy your design directly onto them. You can also use color-copying machines to produce your designs in color. The transparency of copier paper can be used to create many special effects. To achieve a gray or colored tone, simply spray mount a sheet of plain white copier paper onto a sheet of black or colored paper.

Another technique, using the same principle, is creating a screen tint, or "ghosted" background element. To create this effect, separate the artwork to be screened on an overlay on an interim mechanical board (with registration marks) and run this through the copier. Raise the overlay and then make a photocopy of the baseboard. Using a light box, register the two layers of paper together (with the baseboard art as the top sheet) and hinge along the top edge. Raise the top sheet and spray its back with a light coating of spray mount. Adhere the two sheets together and trim to size using the cropmarks as a guide for trimming.

Yet another effect that can be used to indicate gray-colored type is to draw directly on top of the lettering with a white, wax type colored pencil.

Photostats

The fastest and cleanest method to produce an advanced comprehensive is by making a high-contrast photostat. A photostat is made by exposing an interim mechanical, made up of reproduction-quality line and halftone copy, onto light-sensitive photographic paper with a photostat machine (see Chapter 3 under Photostat Machines). The beauty of photostatted comprehensives is that everything is recorded on one surface, as opposed to other methods of producing advanced comprehensives in which type, colors, films, and photographs are adhered to a background in layers.

Photostat machines can be used to create different effects (Fig. 5.4). If you have halftone or continuous-tone copy such as photographs in your design, you can photostat the copy to produce two different results. If you want a black-and-white image without gray tones, adhere the copy directly onto the interim mechanical and photostat it with the line copy. If you wish to retain gray tones, first make a screened photostat, or velox, of the halftone copy, adhere that onto the interim mechanical, and photostat it with the line copy. A velox is made basically the same way as a photostat, except that a dot screen is used in the camera in order to reproduce continuous-tone copy. All elements should be adhered to the illustration board with adhesives such as rubber cement that let you reposition them so that changes can be made. Once the board has been prepared, you can reduce it, enlarge it, or copy it at actual size for presentation or for use in other advanced comprehensive techniques.

The main limitation of photostats is that the type and imagery are indicated in black-and-white. This may be preferred for designs in which dramatic effect is desired and color is not appropriate. For other situations, however, there are methods to add color to the white areas of photostatted comprehensives.

Using fine- or wide-nibbed markers with water-soluble ink, you can apply color in small areas and smooth out streaks with cotton swabs moistened with saliva or water. This technique works well on photostats that are mostly black with small white areas dropped out since colors can overlap onto the black areas without being seen. Photostats that may have many large white areas, such as outlined type or shapes, or line illustrations can also be covered using this method.

Another method of applying color to photostatted comprehensives is with color film. One way is to cut self-adhesive color films to the desired shapes and then burnish them down firmly. You can also apply color film over an entire photostat. Or you can cut the shape you want from the film, using the underlying photostat as a guide, and then cover that area of the stat. The best way to trim the film is, first, to cut the film larger than the size of the area to be covered. Then adhere and burnish the film to the photostat and finally cut the shape you want (with an art

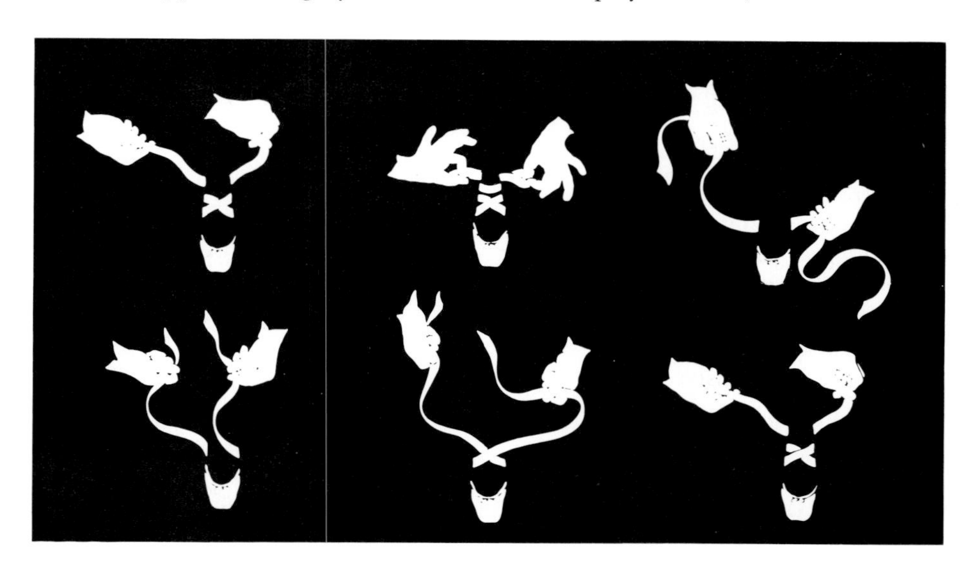

Figure 5.4. These images were drawn at "twice-up" on drafting film with a #2 technical pen and black ink and were then photostatted in reverse to produce a black background with white illustrations. (Artwork: author)

knife) through the film, using the photostat as a guide for cutting. Remove any excess film.

Photostats can also be used to present three-dimensional ideas. You can adhere or wrap photostat paper directly onto a three-dimensional package dummy for presentation purposes (Fig. 5.5). Since photostat paper is thinner than most photographic papers, it can be scored, folded, and trimmed easily. Also, it is durable enough to stand up to glues, cements, and spray adhesives applied to the back.

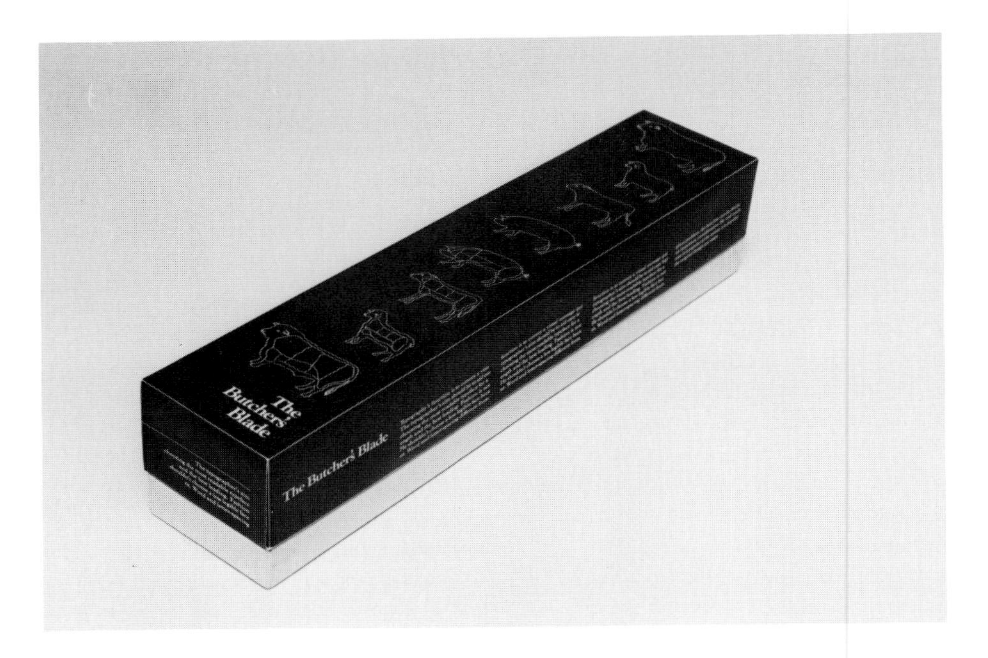

Figure 5.5. A photostat adhered directly onto a set-up box packaging dummy. (Artwork: author)

Photoliths

Another commonly used method of producing advanced comprehensives involves photographing reproduction-quality line and halftone copy with photolith film, such as Kodalith. Photolith film is a high-contrast film on which images and type are exposed with a photostat machine. The design areas in photoliths can be indicated either as positive, with the images and type black and the background clear, or as negative, with the images and type dropped out so that they appear as clear film and the background appears as opaque black (Fig. 5.6).

Both negative and positive photolith films can be used to make advanced comprehensives in color. The simplest method is, first, to adhere colored papers, photographs, foils, or patterns on a piece of illustration board. Place the film over the board and tape it securely in position. Then, cut a window mat and drop it into place to display your piece and to cover the backing board.

After a photolith film has been made of an interim mechanical, you can use it for presentation or for more complex methods of rendering advanced comprehensives.

Figure 5.6. *a.* Positive imaged photolith film over cut Color-Aid paper (Artwork: Scott Osborne). *b.* Negative-imaged photolith film over pink colored paper. (Artwork: author)

Color Keys

The Color Key, manufactured by 3M, serves the same function for presenting advanced comprehensives as photolith film, but it is not limited to black-and-white. The Color Key is a light-sensitive, ink-coated polyester film sheet available in twenty-six Pantone, eleven transparent, and ten opaque colors. It is very durable and will not tear, shrink, or fade. You can use Color Keys to make two- and three-dimensional advanced comprehensives (Fig. 5.7), 35-millimeter slide art, films and videos, transparencies for overhead projectors, and color proofs for printing line and halftone copy. There are also special orange and opaque black Color Keys

Figure 5.7. A Color Key that has been wrapped around a dummy. (Artwork: author)

used to produce film negatives when contacted with positive-imaged photolith films or film positives when contacted with negative-imaged photolith films, which are then exposed with special light-exposing machines. You can use these negative and positive films to make other Color Key colors. Color Keys can also be used in combination with colored papers and films (Fig. 5.8).

To use Color Keys to prepare an advanced comprehensive, follow these steps:

1. Prepare an interim mechanical for Color Keys as discussed in Chapter 4 under Preparing Interim Mechanicals with Several Colors. Remember, if the Color-Key comprehensive is to be more than one color, separate the type and artwork for each color on separate overlays.
2. Make a photolith film negative or positive: If you want the effect of the Color Key to be the same as that on the mechanical, such as having the type and artwork in black and the background in white, make a photolith film negative for each color. If you want the background and foreground relationship to be the opposite of that on the mechanical, make a positive film of each color.
3. After you have made a film of each color, contact or sandwich each sheet separately with a sheet of Color Key and expose each pair to an ultraviolet light source, which will harden the color coating in the image

Figure 5.8. Opaque White Color-Key film over various colored papers. (Artwork: Scott Santoro)

areas. 3M and other companies make special light-exposing machines (see Chapter 3 under Exposure Units), but they are expensive, and there are other methods you can use. One method is to contact the film and Color Key under a sheet of 1/8-inch glass and expose it using a 500-watt photo flood bulb.

4. Coat the exposed film with 3M Color Key developer. Using a Webril Wipe or other nonabrasive wipe, remove color coating in areas in which there are no images.
5. Rinse the Color Key film and blot it dry with paper towel.
6. Place the Color Key over the background material, which can consist of different colored papers, that has been mounted on illustration board.
7. Cut a window mat and drop it over the artwork. If you are assembling many layers of Color Key for the advanced comprehensive, register the sheets and then mat them.
8. For a three-dimensional comprehensive, follow the first five steps, then wrap the Color Key film firmly around the dummy.

If the colors you want are not offered in the Color Key system, you can tint or color by hand—opaque white Color Key film after it has been processed and dried. The 3M company offers a Colorant Kit containing eight tube colors that can be mixed to create virtually any color. In addition, many water-based color media, such as Marvy Markers, can be used to apply color. An advantage of hand coloring is that it lets you combine many colors on a single sheet of opaque white film.

ENHANCING AND PROTECTING COMPREHENSIVES

Finished comprehensives, like finished artwork, should be mounted or covered for enhancement and protection. This section describes various ways of making comprehensives as refined as the finished printed piece, while protecting them in the process.

Varnishing

In reproduction, varnishing is the application of a clear, colorless coating to a printed piece to protect it and enhance its appearance. *Varnish* can be applied to an entire piece or to selected areas (in which case it is called a *spot varnish*.) Two commonly used varnishes, matte and high gloss, are used primarily for aesthetic reasons. It is important to be aware of the fact that the surface texture and color of a paper affect how successful the varnish application will be. For example, when uncoated papers, such as Pantone, are sprayed with a gloss varnish, such as Crystal Clear, they do not take on the intended glossy finish but only absorb the spray and become slightly darker. Coated paper stock, on the other hand, will take on a glossy appearance when sprayed with Crystal Clear. When working with textured coated and uncoated paper stocks, gloss or matte sprays will protect the piece but will not produce a noticeable difference in finish. Similarly, when a light-colored (such as white) high-gloss coated paper is sprayed with a matte varnish, there is usually no visible difference.

Varnish can be used easily and successfully to give a dull effect to glossy paper, thus producing a great contrast between the unvarnished and the varnished areas (Fig. 5.9). To do this, follow these steps:

1. Draw the image to be varnished and its trim size accurately on tracing paper.
2. Apply graphite to the back of the tracing paper, transfer the image, right-reading, onto heavyweight bond paper using the graphite transfer method.
3. Using an art knife, cut away the transferred image to be varnished from the heavyweight bond paper, creating a stencil of the image. To varnish simple shapes, you do not need to use the graphite transfer method. Simply measure and cut the stencil. Next, spray the back of the stencil with a light

a

Figure 5.9. *a.* Letraset matte (protective coating) sprayed over the open area of a stencil *(figure continues).*

b

c

but thorough coating of adhesive and let dry.

4. Turn over the sprayed paper stencil so that the sticky side is facing down, and adhere it flat onto the paper that is to be varnished.
5. Apply several even coats of Letraset Matte (a protective coating) over the open area of the stencil until the paper no longer appears glossy.
6. After the spray has dulled the designated area and has dried, remove the stencil carefully so as not to damage the underlying paper's surface. To help remove the stencil, apply a small, controlled stream of rubber cement thinner.
7. Remove any excess spray adhesive with a cotton swab moistened with rubber cement thinner. You can now use the varnished sheet as a background or base on which to apply other media or type.

Cellophane Wrapping

On some occasions, especially if you are doing a comprehensive for a package, you may wish to simulate a commercially applied thin cellophane wrapping. Such wrapping can be used for protection, enhancement, or both (Fig. 5.10). Use the wrapping technique presented in the following steps to wrap any square or rectangular packaging comprehensive (see also Fig. 5.11).

1. Gather the materials needed for wrapping: a roll of No. 88 or No. 15 cellophane, a No. 0

Figure 5.9 *(Continued). b.* A dull-varnished effect on a glossy paper. *c.* Here the logo is indicated in a subtle manner on a brochure.

Figure 5.10. A commercially applied thin cellophane wrapping is convincingly simulated on this packaging comprehensive.

inexpensive watercolor brush, and a small bottle or can of acetone.

2. Measure the dimensions of the package to determine how much cellophane is needed to wrap the entire piece. This is often a trial-and-error process, similar to gift wrapping. Next, cut a piece of cellophane large enough to wrap the package.
3. Place the cut sheet of cellophane on a clean black mat board so you can see if dirt or dust particles have adhered to its surface. Then place the package on top of the cellophane sheet.
4. To see if the cellophane has been cut to the correct size, wrap it around the entire package. Make any necessary adjustments, such as cutting off excess overlap if the sheet is too large or cutting a fresh sheet if it is too small.
5. Using the properly cut cellophane sheet, tightly wrap the package so that the overlapping seam runs lengthwise. With one hand, apply slight fingertip pressure to hold the seam tightly together. With your other hand, quickly dip the brush into the acetone, wipe it across a clean piece of mat board to remove any excess acetone, and "paint" the entire length of the seam. Be sure to

a

Figure 5.11. Cellophane wrapping. *a.* Cellophane is placed on a clean black mat board so that any dust or dirt particles can be seen and removed before wrapping. Acetone is "painted" along the entire length of the seam while the overlapping cellophane is held tightly together. *b.* Close up the end of the package in which all the seams have been joined, and the bond is set. *c.* Once all of the seams have been joined and the bond is set, the wrapped package is ready for presentation.

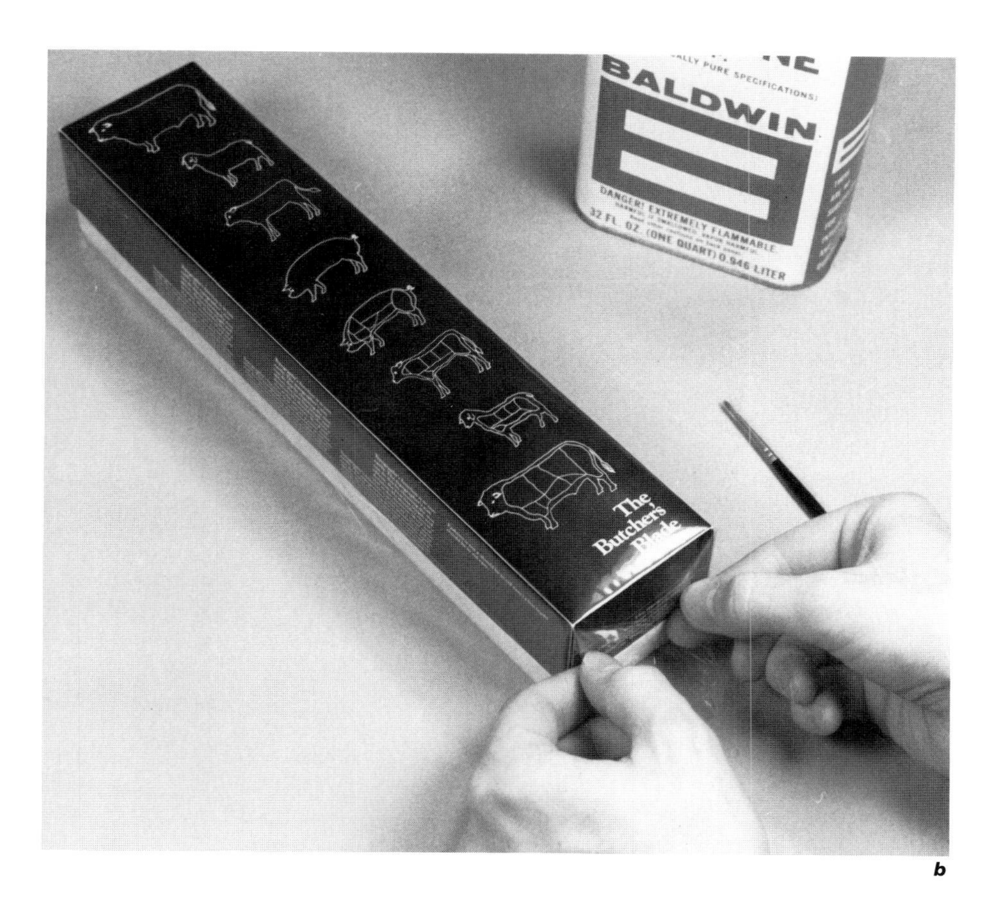

b

c

maintain pressure on the seam until the excess liquid evaporates and the bond is set.

6. On one end of the package, fold the excess cellophane in a manner similar to that used to gift wrap a present. Then bond the overlapping cellophane to itself using the technique described in step 5.

7. Tightly wrap the cellophane around the remaining end, and bond it firmly in place using the technique described in step 5.

8. Examine the bond on all the seams. Touch up any loose spots.

Acetate Wrapping

After you have matted or mounted a two-dimensional comprehensive for presentation, you can further enhance and protect it by wrapping it with acetate.

To wrap a mounted or matted two-dimensional comprehensive with acetate, follow these steps:

1. Cut a piece of 0.003mm-thick clear acetate approximately 1½ inches larger on all four sides than the board you want to wrap.

2. Center the board face down on the acetate. Apply a 1-inch strip of white tape to the middle of each side to hold the acetate to the board.

3. To eliminate overlap when the acetate is wrapped, trim the tip of one corner of the acetate at approximately a 45-degree angle. Repeat this process for the remaining three corners.

4. Remove the tape from the middle of one side, fold and pull the acetate toward the center of the board, and tape it down securely (Fig. 5.12a). Repeat this process on the opposite

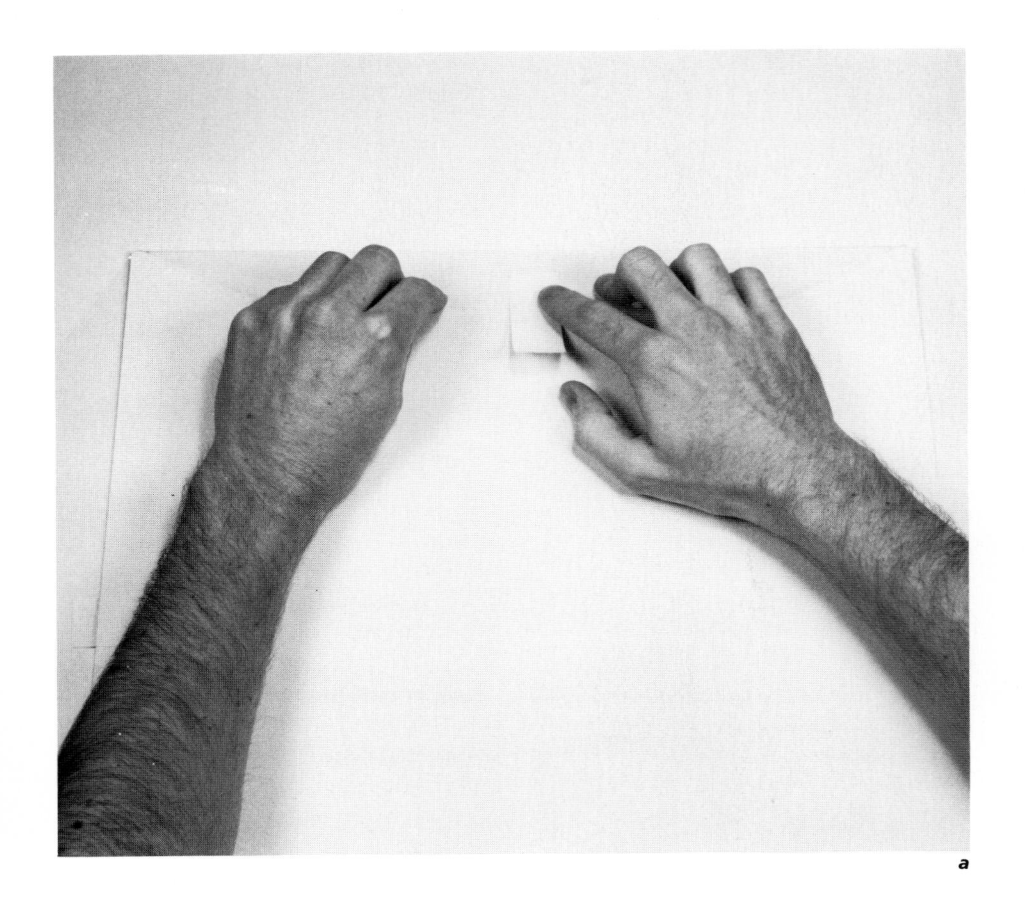

a

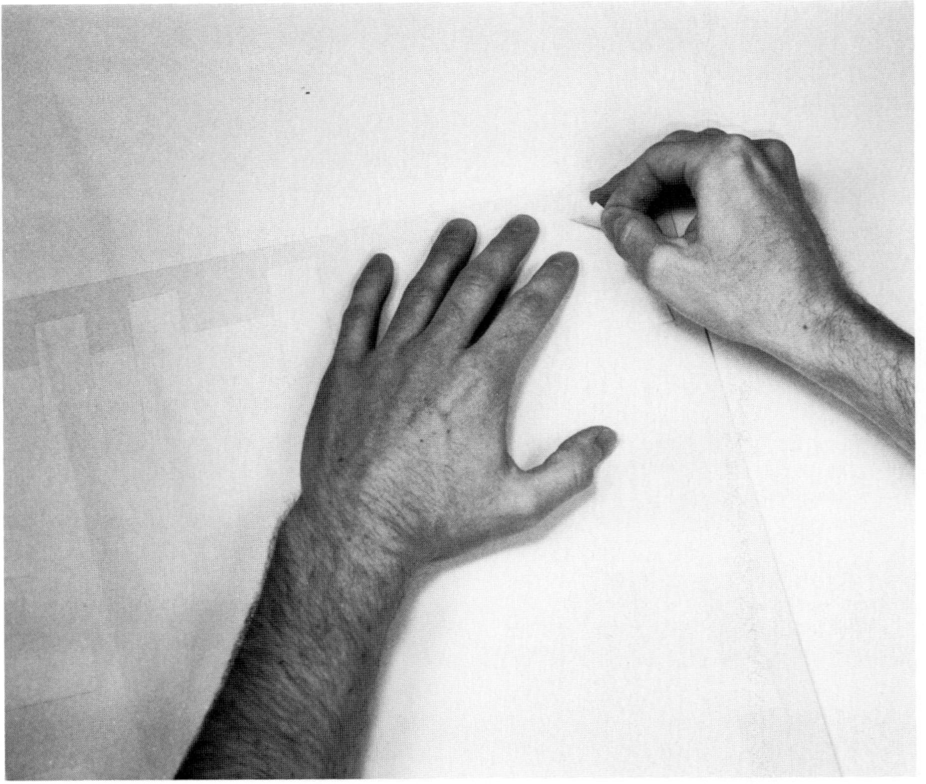

b

side and then on the remaining two sides, pulling tightly each time.

5. Starting from the center piece of the tape on one side, grip the acetate and pull it toward the middle with one hand while affixing a 1-inch strip of tape at approximately 1-inch intervals with the other hand. Tape the entire side, working from the center toward the outside edges (Fig. 5.12b). Repeat this process on the opposite side and then on the remaining two sides.
6. Cut a sheet of inexpensive smooth, black cover stock approximately 1⁄16 inch smaller on all sides than the original board.
7. Apply double-sided tape along the edges of all four sides of the cover stock. Then make a criss-cross through the middle of the sheet (Fig. 5.12c). Turn the taped cover stock over and adhere it securely to the back of the mounting board (Fig. 5.12d).

Figure 5.13 shows the finished acetate-wrapped two-dimensional advanced comprehensive.

Figure 5.12. Acetate wrapping. *a.* After removing the tape from one side, the acetate "flap" is pulled toward the center of the board and is then taped down securely. *b.* While the acetate is being pulled toward the center, tape is affixed at approximately one-inch intervals working from the center out toward the outside edges. *c.* Double-sided tape is applied along the edge of all four sides of the precut cover stock and is then criss-crossed through its middle. *d.* The taped cover stock is then adhered securely to the back of the mounted piece.

Figure 5.13 *(right).* The finished "acetate-wrapped" two-dimensional advanced comprehensive.

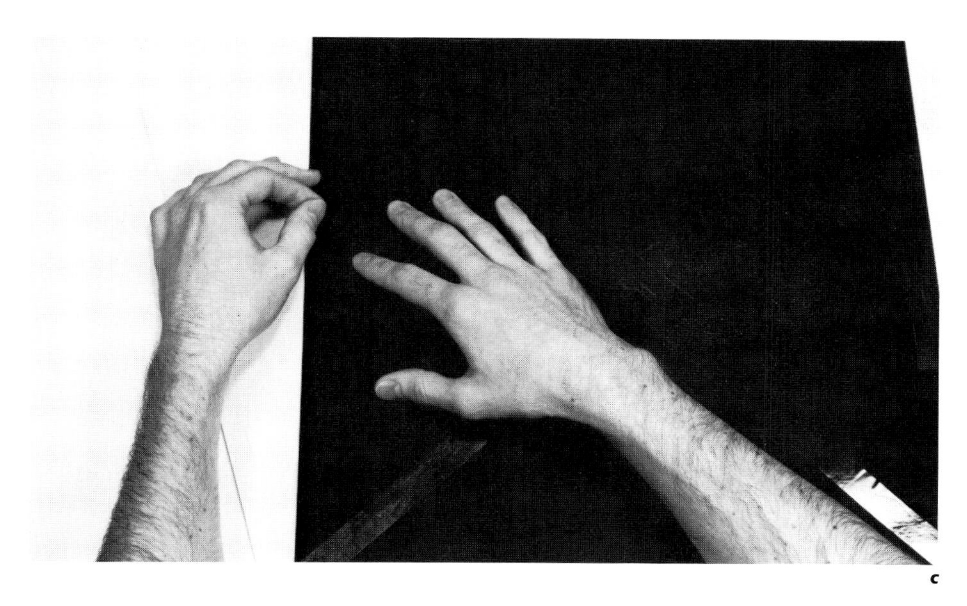

c

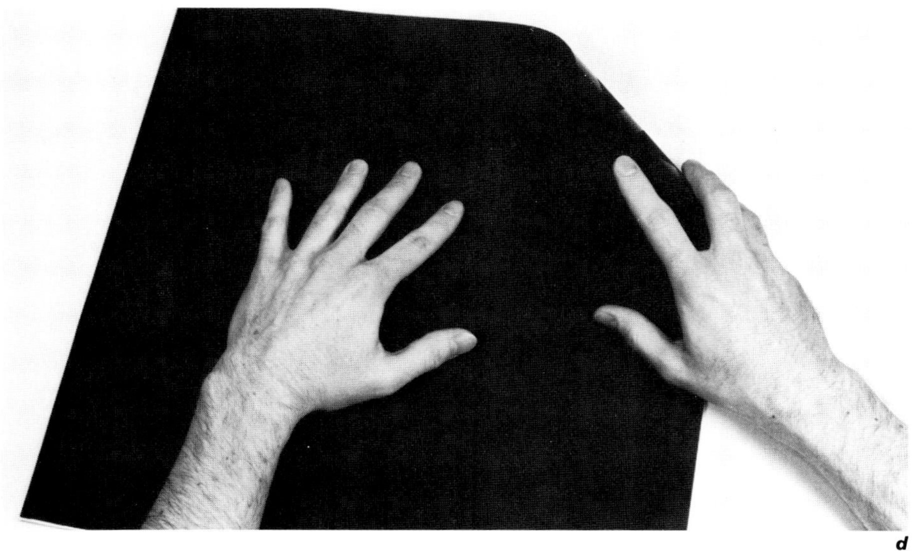

d

SIMULATING PRINTING TECHNIQUES

This section explains various printing processes that can be simulated in comprehensive form. Some involve good hand-rendering skills, and others require expensive materials and equipment. These techniques will make a good idea and layout more believable but will not salvage a weak idea.

Embossing

One printing technique that can be simulated in the graphic design studio is *embossing*. Embossing is the creation of a raised surface image, or relief, on paper or other material (Fig. 5.14). You can emboss an image as a high, low, or multilayered relief, depending on the desired effect. For example, the relief can be one level or many levels of flat planes, or it can be modeled, such as the figures that are illustrated on coins.

Embossing is often combined with a printed ink or metallic foil in register. Or it can be done on unprinted paper, producing an effect called *blind embossing* (Fig. 5.15).

A related effect is *debossing*, which results in a recessed image. The only drawback to this effect is that the tool marks are visible on the front of the paper.

You can create embossed and debossed images quickly and easily by following these steps:

1. Draw the image to be embossed, including its outside dimensions, or the trimmed final piece, on tracing paper.
2. Cut an illustration board approximately 1 inch larger than the outside dimensions of the trim size of the finished piece image. Cut a piece of two- or three-ply bristol board (depending on the desired depth of the embossing) about ½ inch larger than the dimen-

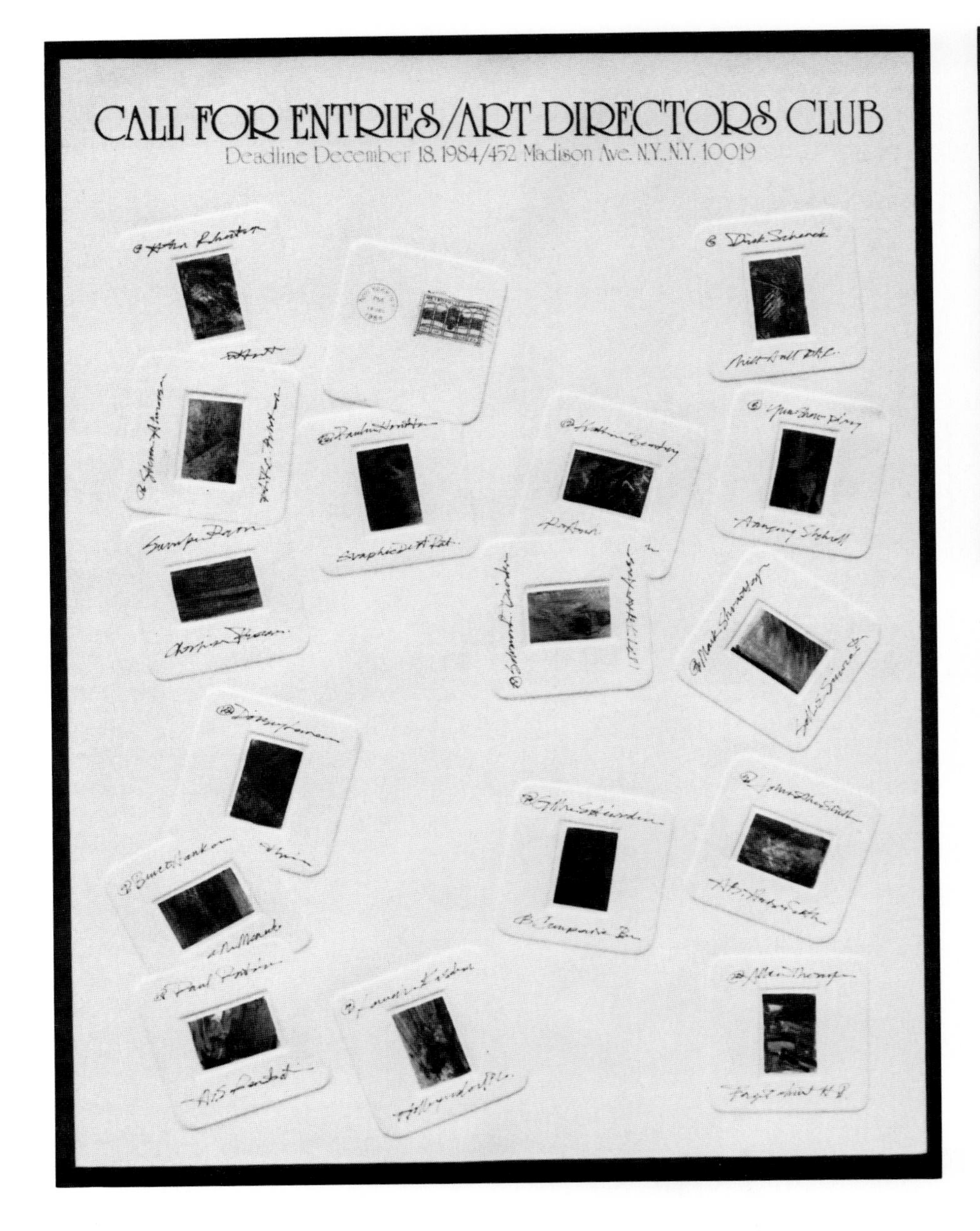

Figure 5.14. A comprehensive layout that simulates embossing. Although special printing techniques such as embossing add beauty to almost any piece, they can also add substantial costs to the total production of a job. (Artwork: Katrina Beasely)

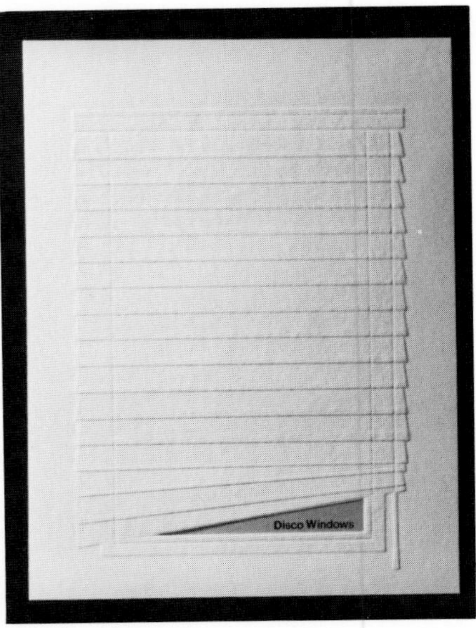

Figure 5.15. This piece combines blind embossing with a simulated die cut. In reproduction, die-cutting is the use of sharp, steel rules to cut special shapes from printed sheets. (Artwork: Bruce Hanke)

sions of the image to be embossed.

3. Coat one side of the bristol board with rubber cement or spray adhesive and adhere it to the center of the illustration board.

4. Make a wrong-reading tracing-paper layout of the image to be embossed and the trim size of the finished piece. Apply graphite to the back using the graphite transfer method discussed in Chapter 4.

5. Hinge the tissue-paper layout along its top edge over the bristol board. Transfer the wrong-reading image onto the bristol board to provide guidelines for cutting.

6. With an art knife and a No. 11 or No. 16 blade, cut the image out of the bristol board and remove the cut-out piece (Fig. 5.16). You can use rubber cement thinner to loosen the adhesive bond. Remove excess rubber cement with a rubber-cement pickup or cotton swab soaked in rubber-cement thinner.

7. Cut the paper to be embossed slightly larger than its final trim size and tape it face down on the bristol board. Since you will have to burnish the image, use a paper that can be stretched and creased without tearing, such as one-ply bristol, heavyweight cotton content text paper, or 65-pound cover stock.

8. Reapply graphite in the image area on the back of the tissue paper. Transfer the image and

Figure 5.16. Removing the excess bristol board from the embossing stencil.

its outlines to the paper to be embossed. These pencil marks will be used as guidelines for burnishing and can be erased after the image has been embossed.

9. Using a fine, round-tipped burnisher, apply pressure on the tip and trace the inside edges of all the guidelines. This will sharply crease the paper at the top of the cut edges of the bristol board (embossing die) for an accurate embossed effect on the opposite side (Fig. 5.17). If the paper you are using has a texture, you should emboss the entire area within the guidelines to flatten the texture so that it is uniform. To check if you have embossed the entire image, remove the lower pieces of tape and lift the tracing paper. When the entire image has been embossed, erase the guidelines.

10. Remove the tape strips and turn the piece over to evaluate the embossed right-reading image (Fig. 5.18). Type and artwork should be rendered after the embossing to ensure correct registration since the embossing itself is done on the back of the paper. Finally, add type and artwork and then trim the piece for presentation (Fig. 5.19).

Another way in which to simulate an embossed effect is by layering one or more sheets of paper that have been cut in the shape of the image to be embossed. Although this technique is not as convincing as the one just described, it is faster.

Figure 5.17. To crease the paper sharply, apply pressure using a burnisher and trace the inside edges of all the guidelines.

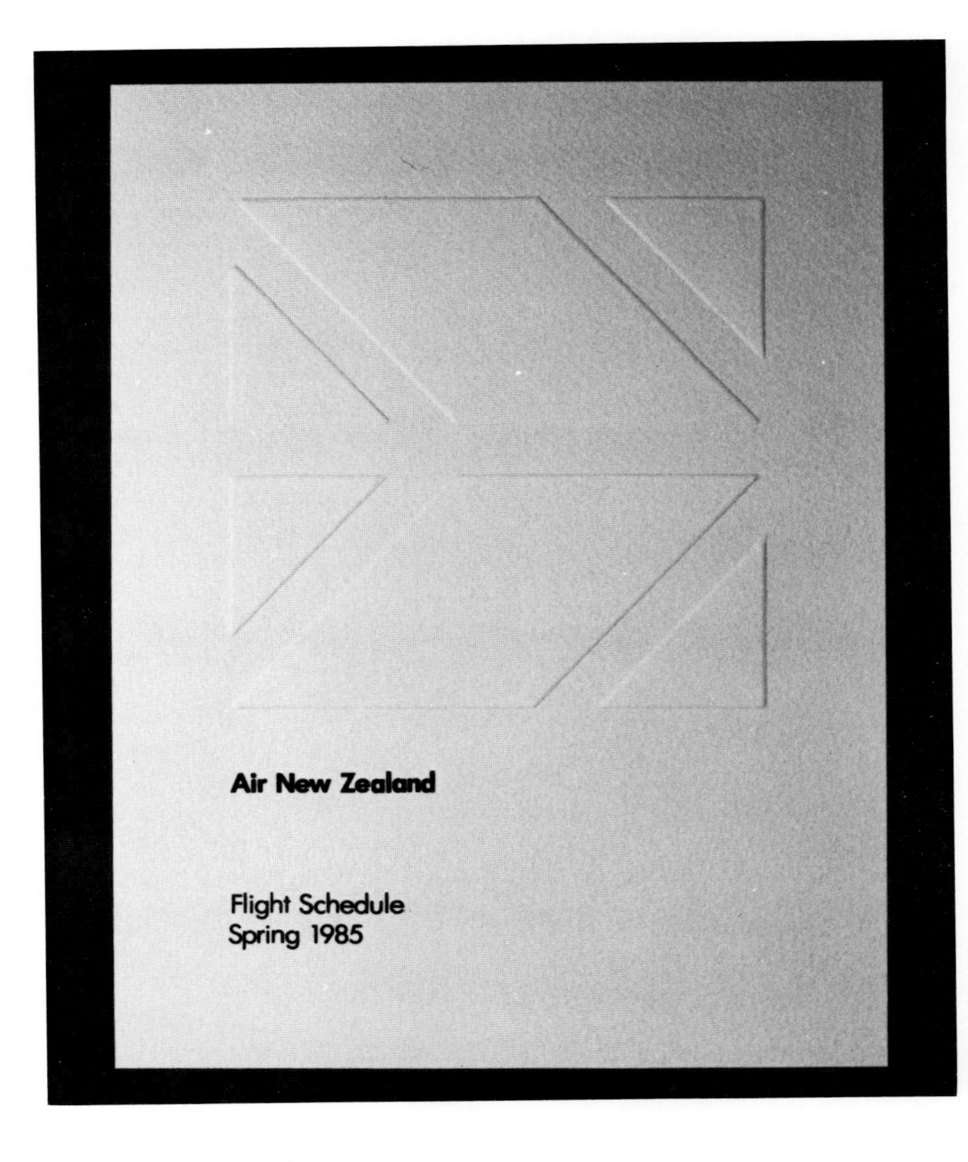

1. Use thin, text-weight paper such as that used for stationery. Cut out the image to be embossed from one or more sheets of paper. For a high image, rubber-cement several sheets of paper together and then cut out the image.
2. Adhere the cut-out image, either from one sheet or from several sheets, onto a new sheet of the paper.

3M Image 'N Transfer System

An imaging transfer system is a process in which type, a logo, or artwork is reproduced in color on a clear or translucent carrier sheet. Some imaging transfer systems use an adhesive backing applied behind the image area, which allows you to burnish down that image, in position on most surfaces. In the advanced comprehensive in Figure 5.20, black and red I.N.T.s were burnished directly onto an airbrushed background.

The 3M Image 'N Transfer (I.N.T.) system, which lets you reproduce and transfer symbols, logos, line and halftone artwork, slogans, and other elements, has several advantages over other film-backed imaging transfer systems (see Fig. 5.21). With other systems, such as the 3M Color Key, the image is permanently etched onto the film's surface, the film adds surface glare that can change the color and texture of the backgrounds, and the film has to be mounted between a window mat and backing board for support, making it quite heavy. With the I.N.T. system, you can burnish I.N.T. transfers from film-backing sheets onto the comprehensive and discard

Figure 5.18 *(top).* Remove the tape strips, turn the piece over, and evaluate the embossing.

Figure 5.19 *(bottom).* Embossed piece ready for presentation.

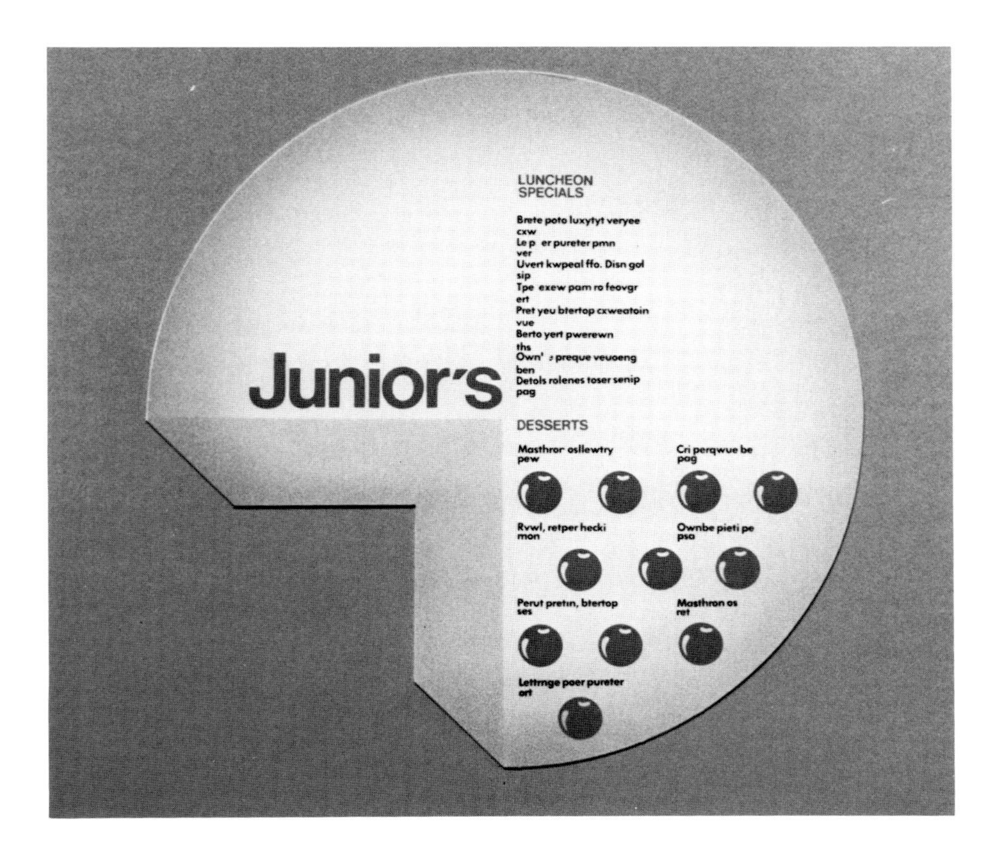

Figure 5.20. This advanced comprehensive combines the use of black and red I.N.T.'s burnished directly onto an airbrushed background. (Artwork: Michael Gerbino)

the excess film, making it unnecessary to create a window mat to hold excess film and eliminate glare.

To use the I.N.T. system, follow these steps:

1. Prepare an interim mechanical for I.N.T. as discussed in Chapter 4.
2. Make a photolith film negative or positive: If you want the effect to be the same as that on the mechanical, such as having the type and artwork in black, make a photolith film negative for each color. If you want the background and foreground relationship to be the opposite of that on the mechanical, make a positive film of each color.
3. After you have made a film of each color, contact each film individually with a sheet of I.N.T. and expose the sheets to an ultraviolet light source, which will harden the color coating in the image areas. 3M and other companies make special light-exposing machines (see Chapter 3 under Exposure Units), but they are expensive and there are other methods you can use. One is to contact the film and I.N.T. under a sheet of ⅛ inch glass and expose it using a 500-watt photo flood bulb.
4. Coat the exposed film with 3M I.N.T. developer. Using a Webril Wipe or other nonabrasive wipe, remove the adhesive-backed color coating in areas in

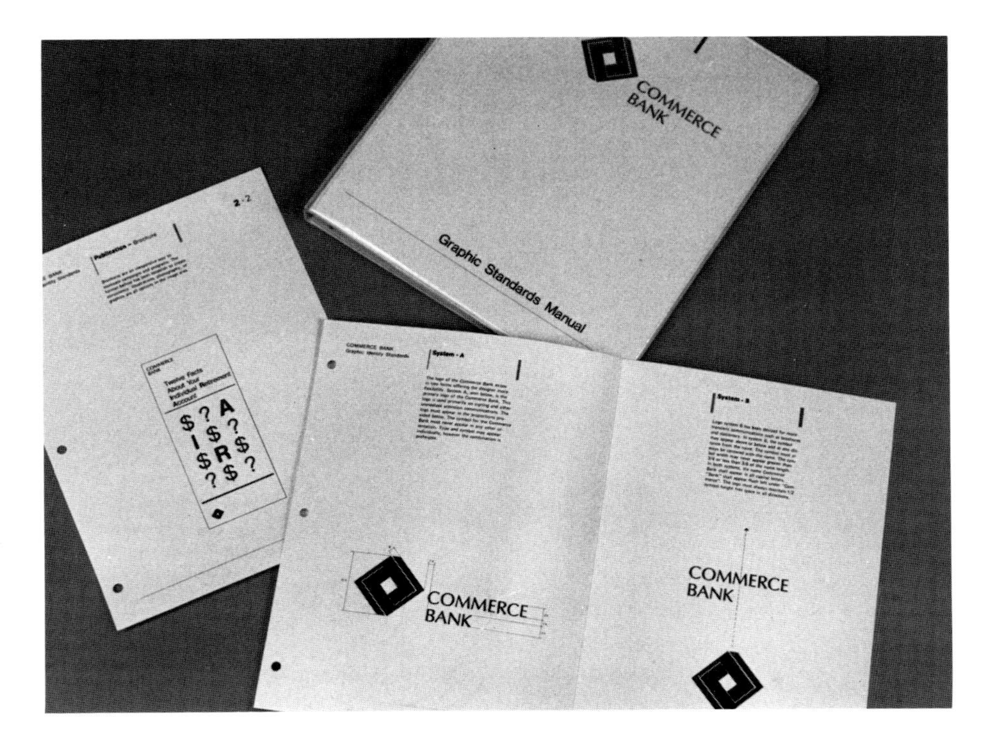

Figure 5.21. Some advantages that rub-down transfers offer over I.N.T.'s is that they can be produced in any Pantone or specially matched color, they provide high resolution of even small serif type, and they can be transformed quickly with any hand burnisher. (Artwork: Randy Tibbott)

which there are no images. The adhesive side will be on the back of the letters if exposed correctly.

5. Rinse the I.N.T. Let it air dry, or dry it with a portable hair dryer turned to the coolest setting.
6. Place a wax paper slip sheet under the I.N.T. carrier sheet to avoid the accidental transferring of images, while positioning I.N.T.'s over the background material. Once in position, remove the slip sheet and transfer the I.N.T. images to your background material. The best tool to use is the 3M burnisher described in Chapter 3, but it is expensive. You can also use less expensive wood, metal, or plastic manual burnishers, also described in Chapter 3.
7. After you have applied the I.N.T. images to the background material, you can mat, mount, or photograph the piece for presentation.
8. Under most conditions, I.N.T. images will not flake, peel, or crack and are therefore suited for three-dimensional comprehensives. Just follow the first six steps; then wrap the I.N.T. firmly around the dummy.

In addition to using the 3M Image 'N Transfer system to make transferable type and imagery, you can use it to create unusual effects.

Opaquing Color Keys

One special effect is created by using I.N.T.'s to overlay a 3M Color Key or dark patterns such as rules, dots, or textures (see Fig. 5.22). Since a Color Key may not be opaque enough to block out a dark, bold underlying pattern, you can back the semitransparent Color Key film with a white 3M I.N.T. transfer. White will

a

b

Figure 5.22 *a.* A 3M I.N.T. was used as an opaque backing for a 3M Color Key. *b.* It was then placed over a dark patterned background.

block out the underlying pattern as well as maintain its overall color value.

To make an I.N.T. to opaque Color Keys, follow these steps:

1. Make a negative photolith film of the interim mechanical, and place it, wrong-reading side up, on top of a sheet of white 3M I.N.T. transfer. Expose the I.N.T. and photolith film for approximately one minute, then develop and rinse it and allow it to dry. The "adhesive" side will now be on the front side of the letters instead of on the back.
2. Tape the Color Key that is to be backed, front side down, securely onto a clean working surface. Position the I.N.T. film over the "wrong-reading" image on the Color Key's back, press it slightly to tack it into place, and tape it onto the Color Key and/or working surface.
3. Burnish the wrong-reading I.N.T. onto the back of the Color Key film with a hand or electrically operated burnisher (see Chapter 3 under Burnishers) until the image is completely transferred from its acetate backing sheet.
4. Turn the Color Key over so that its image is right-reading, and hinge it over or adhere it to its dark underlying pattern.

Color Airbrushing

You can also create unusual effects, such as gradations, rainbows, and tints, by combining the I.N.T. process with airbrushing and its unlimited range of colors. For example, you can create gradated colored type on a solid white or colored background or type that drops out from a solid colored background, or produce colored type on a colored background. The following is the technique for producing colored type on a white or light colored background.

The best paper to use for this procedure is a nonglossy semiporous paper such as bristol, which allows the dye but not the I.N.T. to penetrate the surface. The recommended color media are Dr. Martin dyes (see Chapter 3). Their consistency makes them suitable for use with airbrushes, they can penetrate semiporous paper, and they are compatible with transferred I.N.T.'s. In this procedure, the I.N.T. is used as a *frisket,* or mask, to prevent the background from being sprayed with color (see the first four steps).

1. Expose a positive photolith film (black type on a clear background) with a sheet of white I.N.T. for approximately one minute. Note: This is the reversal of the technique described earlier, which provided you with an I.N.T. that maintained the same negative/positive relationship as found on the mechanical.
2. Develop the I.N.T., rinse it, and let it dry. When it is thoroughly dry, trim the white I.N.T. to a size that leaves about a one-half-inch border around the type, place the film over the paper being used for the comprehensive, and tape it securely into position.
3. Burnish the I.N.T. onto the paper until the entire image is transferred. Remove the plastic transfer carrier sheet and discard it.
4. Cut a frisket out of tracing paper and tape it onto the transferred I.N.T. with white tape. Make sure that the type area is unmasked but that the entire background is protected (Fig. 5.23).

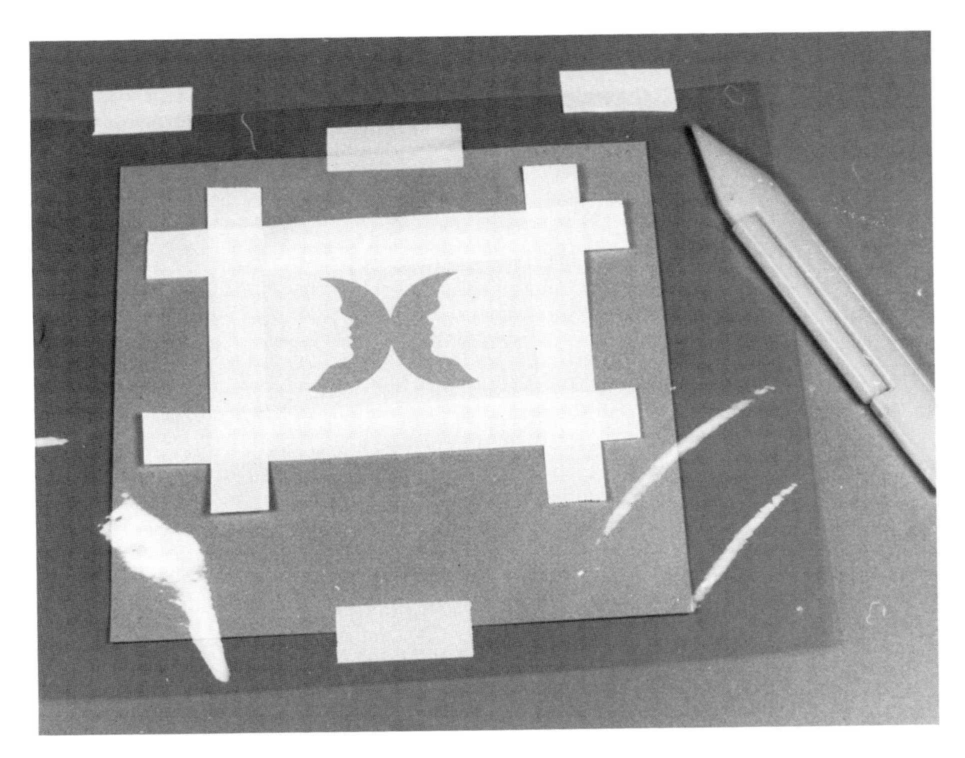

Figure 5.23. An I.N.T. used as a frisket for airbrushing. Tracing paper is taped onto the I.N.T. in order to protect the background from any overspray.

5. Using an airbrush and Dr. Martin dyes, spray the unmasked type area until its surface is coated adequately. Repeat the process if further coats or additional colors are desired.
6. After the dye has dried, remove the tracing paper and I.N.T. mask. To do so, soak a clean cotton ball with rubber-cement thinner until it is thoroughly moistened. Rub the cotton over the entire surface of the I.N.T. to loosen its bond. Soon the I.N.T. will begin to break up and you can wipe it away easily. Remove any stubborn pieces of I.N.T. with a cotton swab soaked with rubber-cement thinner or with a small strip of white tape (Fig. 5.24).
7. After the I.N.T. is completely removed, mount and present the sheet with the rendered type in color as is or use it as base art for other comprehensive procedures (Fig. 5.25).

Figure 5.24. White tape and rubber-cement thinner are used for the removal of stubborn, remaining pieces of I.N.T.

Rub-downs and the Chromatec Process

Rub-down transfers, which can be made from black-and-white line art prepared as an interim mechanical, are based on the same principle as I.N.T.'s in that they are opaque images that are transferred from a backing sheet, using a burnisher, onto a background. These transfers offer several advantages. They have greater clarity, are easier to apply, and are available in any Pantone or special match color. They have two main disadvantages, however, which prohibit their use in most situations: They require very expensive equipment, and the process used to make them cannot be purchased by students or most design studios but are available as a specialized service to any designer who needs to produce superior-quality advanced comprehensives. As a cost-saving measure,

Figure 5.25. When applied properly, this method consistently provides clean, tightly rendered type and artwork.

gang up type and artwork for one or more jobs as shown in Figure 5.26.

There are three popular methods used to produce rub-down transfers: direct-screen printed images with adhesive backing, the Identicolor system, and the Chromatec system. The Chromatec system is the only method that can be used to produce rub-downs in the design studio, so the following discussion will be restricted to that technique. Figure 5.27 shows the equipment used in the Chromatec system. Keep in mind, however, that the equipment is expensive and the process is time-consuming.

The Chromatec transfer system lets you make custom color dry-transfer sheets from a photolith film negative. Chromatec transfers are excellent for making reproductions of posters, annual reports, packaging, signs, logos, and type that look as realistic as the final printed piece. This system offers a wide range of colors, including transparent, opaque, fluorescent, and metallic colors. The dry-transfer sheets have excellent adhesion, are heat-resistant, and can be transferred directly onto most surfaces without peeling or cracking.

To use the Chromatec system, follow these steps:

1. Prepare an interim mechanical for the Chromatec system as discussed in Chapter 4. If more than one color will be used for the Chromatec comprehensive, place line and halftone copy for each color on a separate overlay. Put registration marks on the baseboard and add acetate overlays for accurate positioning of overlays when you assemble the comp.
2. Make a photolith film positive through all the overlays, or colors. This will be used as a guide to position rub-down elements on the comprehensive. To maintain the same black-and-white relationship as found on the mechanical when the

Figure 5.26. Art and type for several jobs have been assembled on one board for a rub-down transfer sheet. A Pantone color chip for the desired color is adhered onto the board along with any written instructions. The dimensions of the transfer sheet should be indicated in pencil on this board.

Figure 5.27. The Chromatec transfer system for producing rub-down transfers in-house.

transfers are produced, also make a photolith film negative for each color.

3. On a clear sheet called Chromaslick, apply Chromatec ink in the desired color. Apply white Chromatec ink on top of the colored ink to make the color opaque. Next, apply a Step II (or residue-free) adhesive. Finally, apply the photocoat. Dry the Chromaslick between each application of ink.

4. Contact the coated Chromaslick with the photolith film and expose them to an ultraviolet light source, which will harden the color coating in the image areas. Chromatec and 3M make special light-exposing machines that will allow you to contact the film and Chromatec under a sheet of ⅛ inch glass which presses the two tightly together during exposure. Always make sure that the film is held under the glass between the light source and the Chromatec sheet during exposure.

5. Rinse the Chromaslick sheet until only your image area remains and let it air-dry.

6. Develop the rinsed Chromaslick using Chromatec developer.

7. Spray the developed sheet with Chromatec's D-3 and remove the photocoat with a Webril Wipe, leaving the exposed adhesive dry.

Silk Screens

Silk-screening, also called *screen printing,* is a printing method in which ink is applied to a stencil adhered to or photographically exposed on a fine-mesh screen. The process, which is clean, accurate, more versatile and permanent than most advanced comprehensive procedures (and less expensive than professional printing methods), is used for short-run promotional pieces, for large runs of comprehensives, and for posters, binders, T-shirts, interior and exterior sign comps, and three-dimensional shapes such as boxes and bottles (Fig. 5.28).

The equipment, tools, and materials needed for silk-screening are inks, screens, stencils, and a press (see Fig. 5.29). Silk-screening inks are available in a wide range of colors and types, including metallic, fluorescent matte, glossy, water-based, and enamel. They are generally dense enough to let you print light colors over dark backgrounds, thereby providing smooth opaque coverage without bleed-through. The inks can be screened onto many different materials, such as plastic, glass, fabrics, wood, metal, cardboard, acetate, foils, and paper, making the process extremely versatile (Fig. 5.30).

The screen consists of a fine-mesh material, such as silk, nylon, Dacron or metal, stretched tightly over a

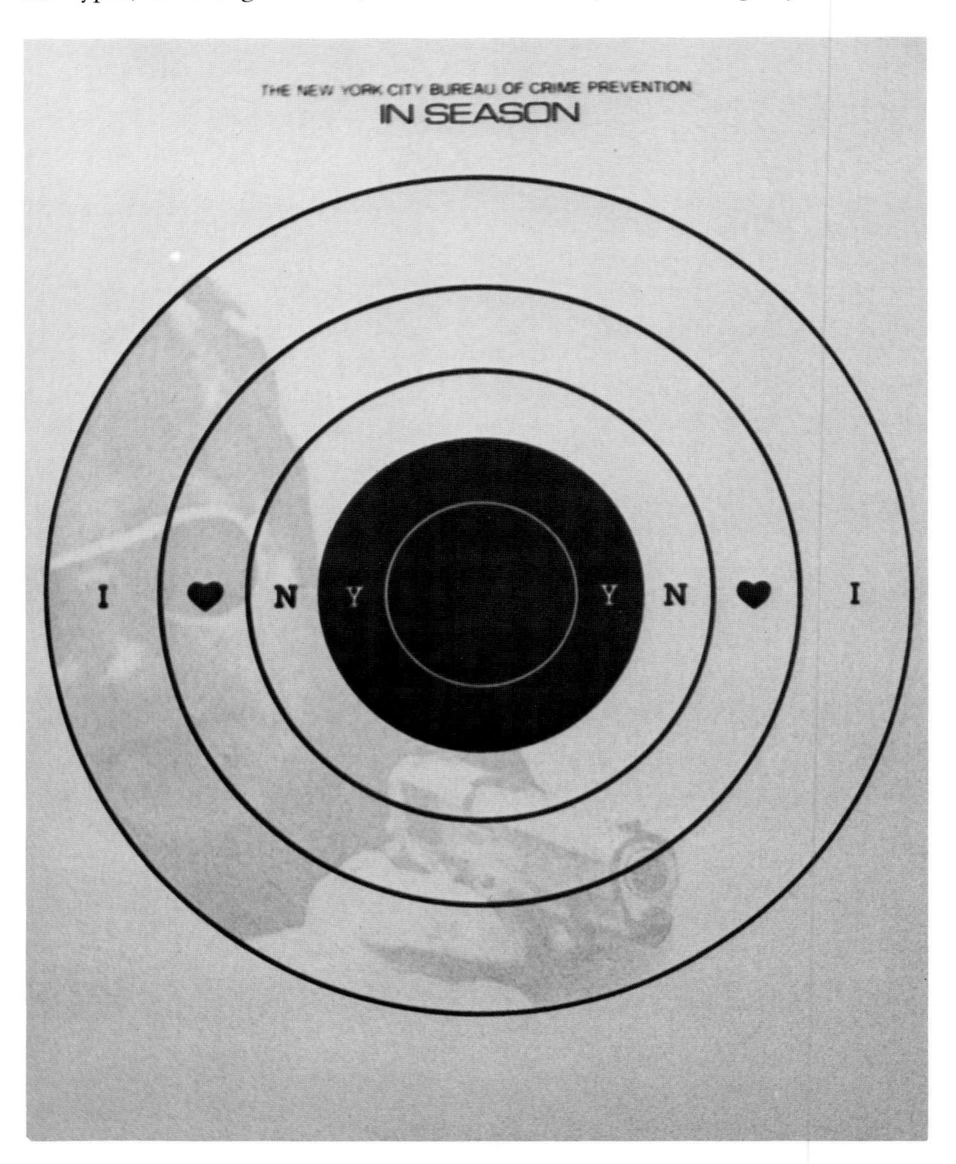

Figure 5.28. Silkscreening-rendered advanced comprehensives that combine both line and halftone copy. (Artwork: Bruce Hanke)

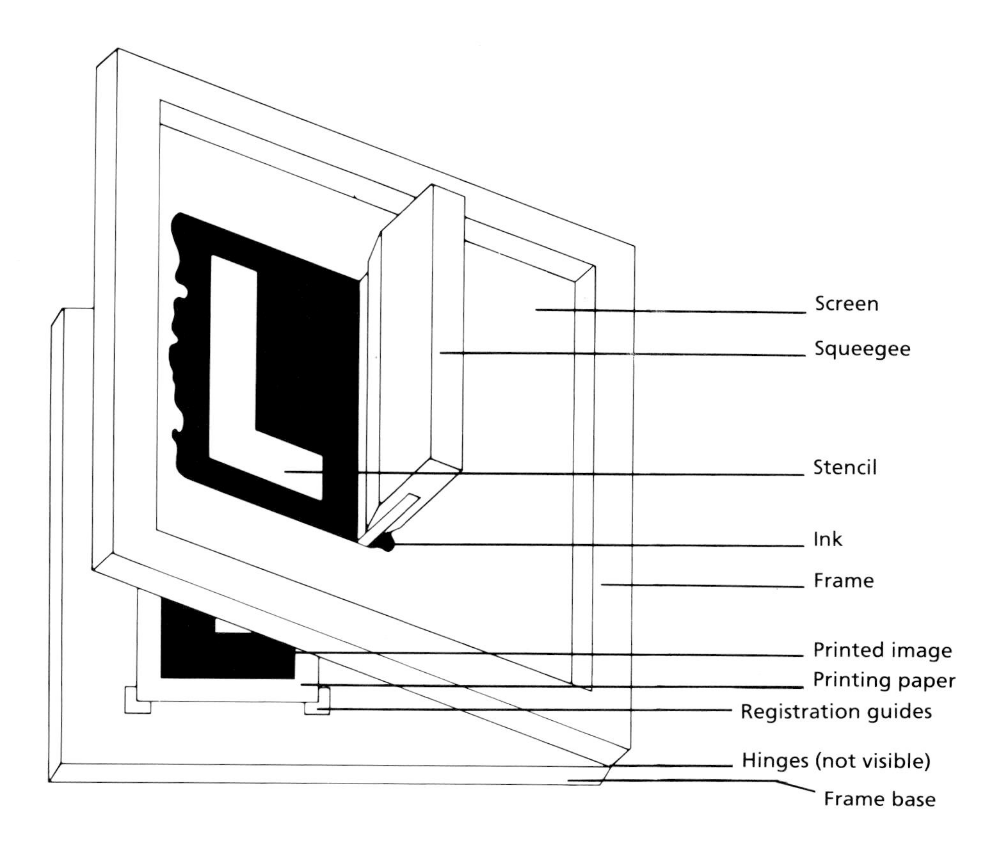

Figure 5.29. Components of a silkscreening unit.

Figure 5.30. This silk-screened book cover would have been impossible to render by any other advanced comprehensive technique. (Artwork: author)

frame. The stencils are films that are cut by hand and adhered to the screen itself or exposed photographically with a photolith film. Hand-cut stencils produce simple and coarse images and are rarely used for advanced comprehensives, unless a rough effect is desired. Silk-screening presses range from inexpensive hand-operated units to expensive fully automatic printing machines.

To photographically expose a stencil, you will need the following materials: a screen unit consisting of a frame with a nylon screen stretched over it and attached by a hinge (which allows removal for easy cleaning) to a background; a vacuum exposure unit or a 500-watt photo flood bulb; a rubber squeegee; ink and ink solvent; and printing paper.

The steps for using this process are as follows:

1. Coat the surface of the screen evenly with a "direct emulsion" developing liquid. Let dry.
2. Expose a photolith film positive (for each color with register marks), right-reading, with the screen using a vacuum exposure unit (which is made for exposing plates for printmaking) or a 500-watt photo flood bulb. If other colors are to be silk-screened, other films can be assembled together on the same screen if time and space permit.
3. Remove the emulsion using a gentle spray of water. Let dry.
4. Lift up the screen frame, slide the printing paper underneath and position it using the image on the screen as a guide. Tape the paper in position on the baseboard.
5. Using a metal spoon, scoop out the thick ink onto the screen above the open image area and apply a generous amount of ink in a straight, horizontal line above the stencil opening,

extending it beyond the opening in both directions. With a squeegee pull the ink toward you so that it covers the stencil opening and is applied to the paper below.

6. Remove the paper. Let the ink dry and then apply type and/or artwork or other silk-screened ink colors if more are to be added to the design. If other colors are to be added, repeat the steps described for each new color.
7. Clean the screen and squeegee with solvent or water, depending on what the ink base is so that the screen can be reused.

Silk-screening is often used to prepare package designs for focus groups in which consumers need to believe they are handling the real thing. It is also used to present clients accurate samples before printing in volume using a more expensive process. Finally, because of the clarity and brilliance of silk-screened package comprehensives, they are often used in print and television advertising.

CREATING SPECIAL EFFECTS

As you become proficient with the skills and materials necessary to produce presentation-quality advanced comprehensives, you might want to experiment with some special effects. In this section, you will learn how to create effects that will differentiate your work from that of others.

Photographic Materials

A special effect in comprehensives can be created by using photographic materials. These can be used as an element in a comprehensive (Fig. 5.31) or as a base on which you can render other comprehensive and advanced comprehensive techniques. If you use photographic materials as a background, you can apply color media, such as gouache or acrylic, directly onto the materials or onto an overlay, depending on the desired effect.

You can create photographic materials from black-and-white or color negatives, slides, or transparencies. You can also clip photos from reference sources such as picture files, books, and magazines. Clipped, colored photographs can be used as they are, or they can be duplicated in black-and-white or color, or be photostatted to form a graphic "posterized" image, without gray, or middle, tones. Photographic images can also be screened to make veloxes or line conversions for reproduction. Some photostat machines can illuminate a slide or transparency from underneath the copy board in order to transform the image into a screened black-and-white paper print.

CREATING A GRID

A grid is a flexible planning aid for laying out a single page, folder, booklet, or package (Fig. 5.32). It helps to create order by aligning text, rules, headings, and artwork.

There are many books on the subject, such as *The Grid* by Allen Hurlburt, that describe in depth how to use grids as well as providing actual grids for standard-sized sheets. In some cases, however, it might be desirable to create a custom-designed grid for a specific need. A step-by-step procedure has been developed to aid in creating a grid for any communication.

Before creating a grid you should decide on a format or outside dimension (if one has not been already determined). Once the format has been chosen, an evaluation should be made of the written and visual information that is to be included in the design. Considerations at this point include the size and number of illustrations and the text matter, including headlines, subheadings, and captions. A series of rough, quickly drawn layouts will usually help to determine which type of grid will help to present the information in the clearest manner.

An 8½-by-11-inch format, seven-column grid has been chosen to illustrate the procedure.

1. Lightly draw, in pencil, a horizontal line across the top of the format that is at least 1½ picas in depth. Draw vertical lines that are also at least 1½ picas from the sides of the format.

Figure 5.31. The logotype was cut out of a color photograph and used as a design element on an advanced comprehensive. (Artwork: Bruce Southard)

2. Draw a horizontal line that is at least 3 picas from the bottom of the page. More space is usually added along the bottom of a format in order to compensate for the "visual" force of gravity.

3. Divide the space inside the margins into equal vertical columns with equal space between each column. In order to compute the width of columns quickly and accurately, first measure the width of the space that the grid modules will occupy. In this example it is a total of 45 picas or 540 points. Note: it is necessary to convert picas to points (which means you need to multiply the number of picas by 12) and convert points to picas (which means dividing the number of points by 12) many times throughout this exercise.

4. Decide on the amount of space between the columns, in this case 1 pica, and multiply this number by the number of spaces between the columns, in this case 6, creating a total of 6 picas or 72 points. Subtract this figure, 72 (in points), from the first figure, 540, to obtain 468. Divide this amount by the number of columns, 7, to obtain the width of each column, 66.8 points. This figure can then be divided by 12 in order to find its pica equivalent (66.8 divided by 12 is 5.57, or roughly 5½ picas).

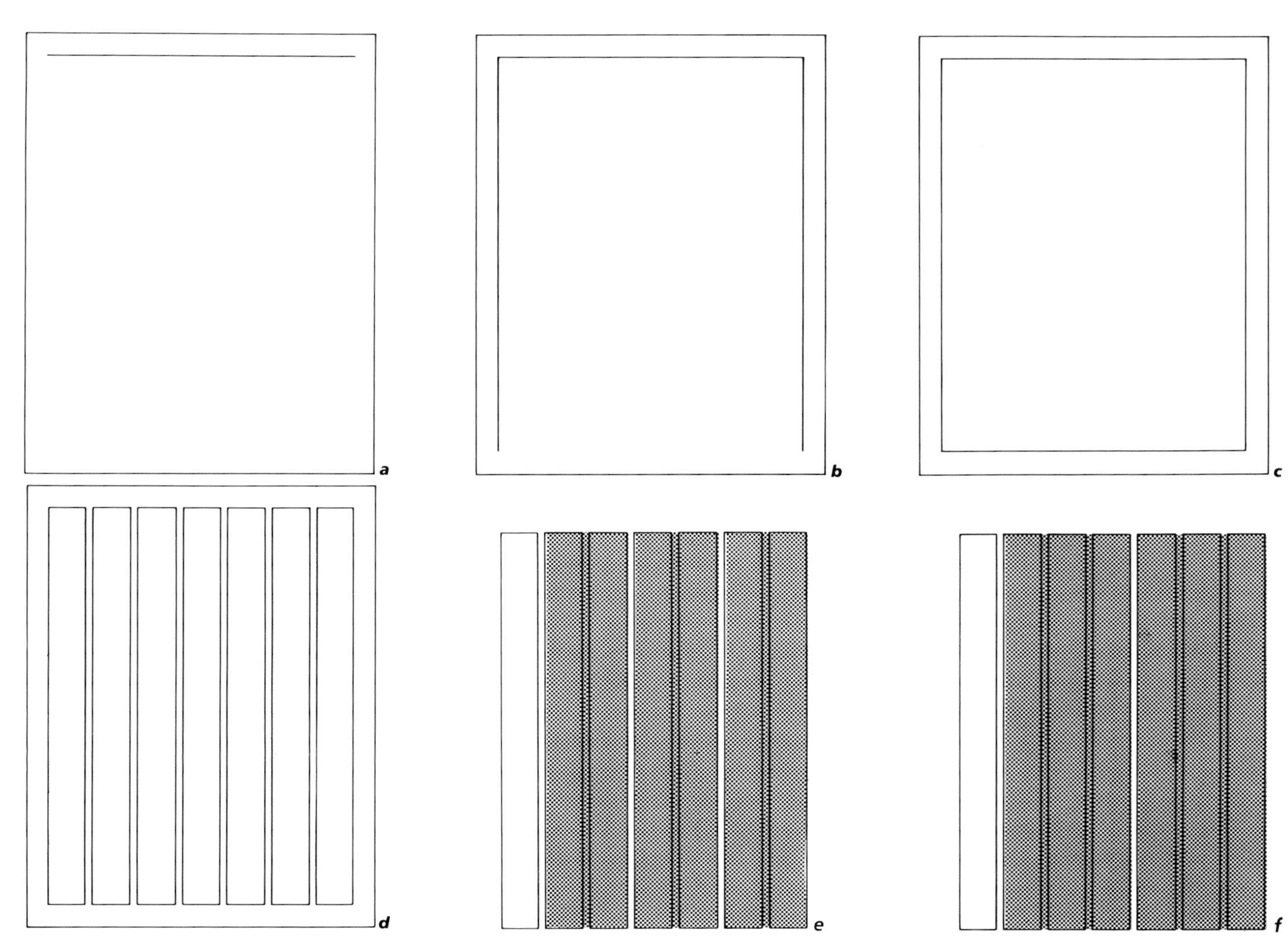

Figure 5.32. Creating a grid. *a.* Draw a horizontal line across the top of the format at least 1½ picas in depth. *b.* Draw vertical lines at least 1½ picas from the sides of the format. *c.* Draw a horizontal line at least 2 picas from the bottom of the page (to compensate for the "visual" force of gravity). *d.* Divide the space inside the margins into equal vertical columns with equal space between each column. *e* and *f.* This seven-column grid consists of seven equal columns with equal space between. They can be combined in many ways as these examples illustrate: two columns of three modules *(e)* and three columns of two modules *(f).*

chapter 6

Presenting Designs

The presentation of your work on each assignment should show as much thought as the design of the finished piece. Even the best ideas can be ruined by a sloppy, poorly planned presentation. An organized, neat presentation tells the person reviewing your work that you care about your work and theirs as well.

The presentation of your work in a portfolio allows the viewer to assess easily the range of assignments you have worked on, your design capabilities, and what kinds of assignments you are suited for. The portfolio serves as a visual résumé of your design work.

There are countless ways to present layouts and comprehensives. Often the choice will depend on whom the work will be presented to. If an idea is being presented to an instructor or another designer, for example, it may not have to be as finished as if it were being presented to a client or a prospective employer. Regardless of how finished the presentation is it must be neat and all the pieces must be consistent in the size and color of the board on which the pieces are mounted (Fig. 6.1). Beginning designers often present their work on mat boards of several different sizes and colors to fit the individual pieces. This approach tends to weaken the impact when the work is viewed collectively.

The challenge in presentation is determining how to enhance your piece without introducing distracting or competing elements (Fig. 6.2). Equally challenging is putting together the best presentation without spending too much time and money.

Figure 6.1. Neatness and consistency are important factors in any type of presentation.

Figure 6.2. The biggest challenge in presentation is knowing how to enhance a piece without introducing a distracting or competing element. This poster was "laminated" (or hot-sealed in plastic) because it was felt that a mat board would be too cumbersome and would only increase its already large size.

CHOOSING THE PIECES TO PRESENT

When you are ready to show your work to prospective clients or employers, you have to decide how many and which pieces to show. Generally, fifteen or twenty matted, mounted, or photographed pieces (see Preparing for Presentation) will represent a good cross section of your capabilities. More pieces indicate a lack of confidence or decisiveness; fewer pieces may not show your capabilities adequately. In addition to the fifteen or twenty mounted, matted, or photographed pieces, you can show your preliminary sketches from one or two of the assignments to illustrate your thinking process and the clarity with which your ideas were conveyed on paper (Fig. 6.3).

PREPARING FOR PRESENTATION

You can prepare your pieces for presentation as you finish each one or all at once when you decide which pieces you will present for a project, to a client, or to a potential employer. In either case, as you prepare each comprehensive or layout, keep in mind that you will probably show it along with other pieces, and try to keep the presentation as consistent as possible. For example, consider the size of the mat board or mounting for a particular piece in relation to your other pieces. For a uniform, professional presentation, you should consider flush-mounting all your two-dimensional pieces on mat board cut to only one to three sizes. Insert three-dimensional pieces such as folders, booklets, and packaging comprehensives in protective sleeves, or photograph them and show them as slides, transparencies, or color prints. This does not mean that the size and shape of your portfolio should dictate the nature of your graphic design work, but only that you think about the overall look of your presentation (Fig. 6.4).

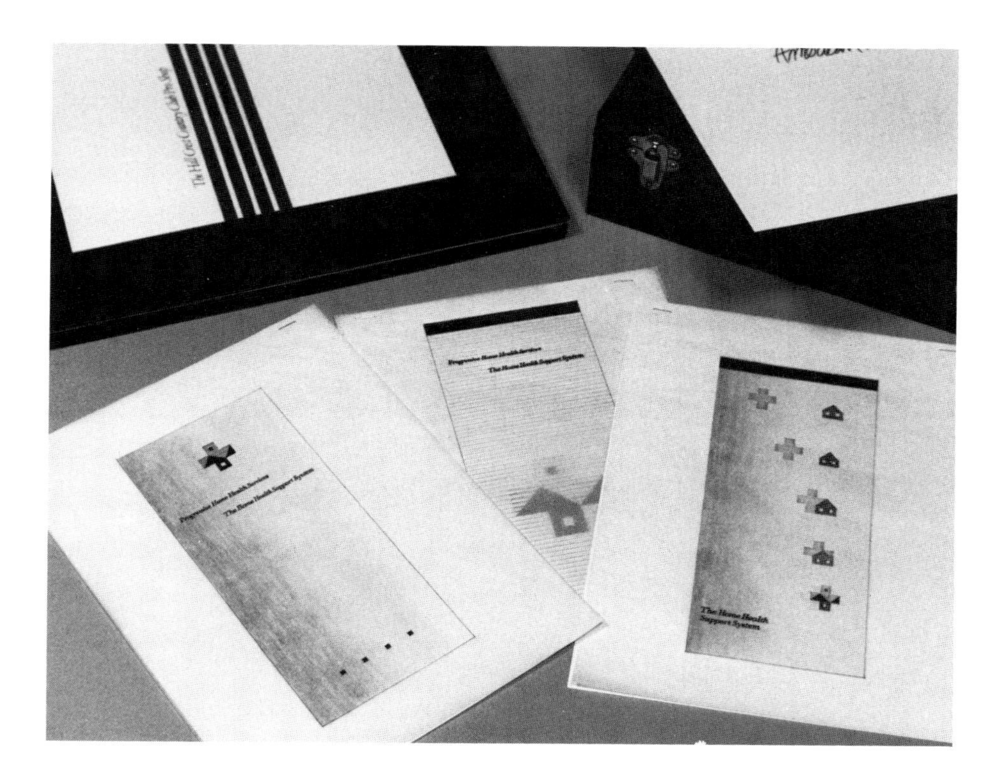

Figure 6.3. In addition to the 15 or 20 pieces that are normally included in a designer's portfolio, preliminary sketches from any one or two assignments should be added in order to illustrate the thinking process involved and the clarity with which the ideas were conveyed on paper.

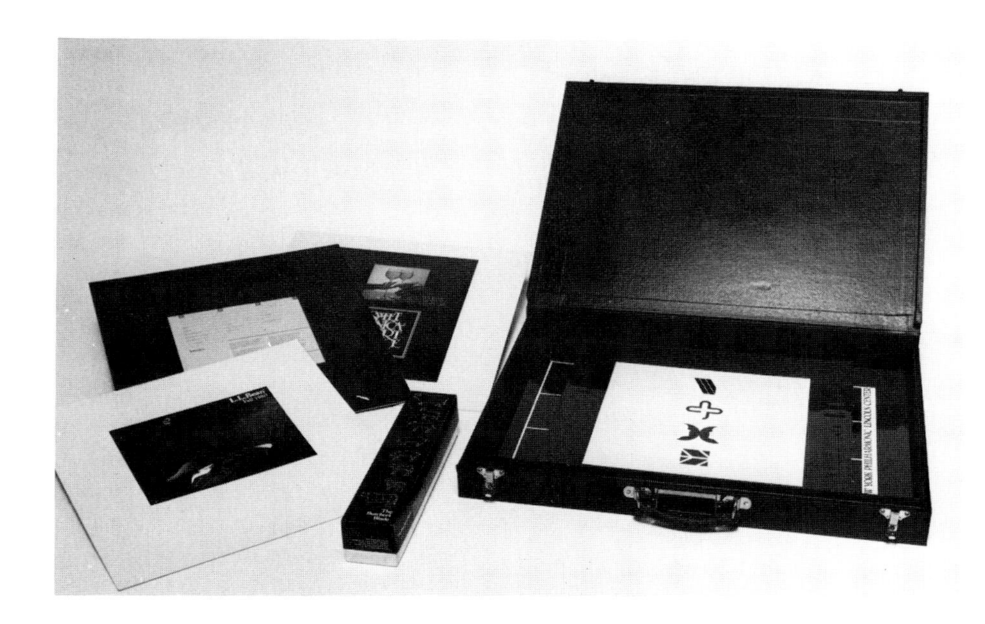

Figure 6.4. Consider the overall look of your presentations.

Two-Dimensional Pieces

A general rule in presentation is if all your work is two-dimensional, it will look neater matted with or mounted on black, white, or gray mat board (Fig. 6.5). If you insert your pieces in a ring binder, however, mount them on the paper that comes with the plastic sleeves or on a suitable substitute.

As an alternative to matting and mounting, some designers use lamination as a presentation medium (Fig. 6.6). Although this process offers professional results, it is expensive and might not be able to be used with some layout and comprehensive rendering media.

Three-Dimensional Pieces

If all of your work consists of three-dimensional and large, two-dimensional pieces, you can present the pieces as they are or record them on 35-millimeter slides or on 4-by-5- or 8-by-10-inch transparencies. If you make slides, insert them into a carousel slide projector or slip them into precut cardboard mounts, which will hold and protect them. Slides can also be placed in three-hole punched acetate sleeves, each of which can store up to twenty slides. The sleeves can then be inserted into three-ring binders for storage and presentation (Fig. 6.7).

If you make color transparencies, you can sandwich them between two same-sized window mat boards for support, protection, and optimum viewing (Fig. 6.8). You can also insert transparencies (depending on their size) into acetate sleeves in a three-ring binder. Another option is to make color prints from film negatives, slides, or transparencies and mount them onto mat board for presentation.

Sketches and Layouts

If you present sketches or layouts with other artwork, their presenta-

Figure 6.5. The choice of whether to mat *(left)* or mount *(right)* is usually determined by the media that were used to render the original piece. For example, the piece on the left was matted simply because it contained a photolith film negative that was placed over a rendered background. This "layering" effect eliminated mounting as a presentation alternative.

Figure 6.6. Lamination, a thin, clear plastic "coating" that is applied with heat to form a hard protective layer over mounted two-dimensional artwork, is used by some designers as an alternative to matting and mounting. The backside of a laminated piece is usually coated with felt in order to prevent scratching.

Figure 6.7. A large volume of slides can be filed in three-hole-punched acetate sleeves and inserted into three-ring binders for easy referral and storage.

Figure 6.8. 4″ x 5″ color transparencies that have been sandwiched between two same-sized window mat boards for support, protection, and optimum viewing. Each image can easily be seen by simply placing the transparency in front of any available light source.

tion will be determined by the media in which you rendered them. A common method for presenting sketches is to show them exactly as they were rendered—as individual sheets torn out of a pad. To present them in this manner, simply slip them into a manila or acetate sleeve, or into a gray clasp envelope.

Two presentation methods will make your sketches more opaque, as well as protect them. One is to cut a window mount out of heavyweight bond paper, then back the sketch with another piece of bond paper cut to the outer dimensions of the mat. Another method is to staple the sketch onto a same-size sheet of heavyweight bond; use only two staples along the top edge to hold the sketch in place (Fig. 6.9).

After a sketch has been prepared, you can photograph it as a slide or photocopy it and insert it into a protective sleeve or envelope. You can also make photocopies on three-hole punched paper and insert them into a binder for protection, storage, and uniform presentation (Fig. 6.10).

DETERMINING THE FORMAT

When you make a presentation to an instructor or client, you usually show your pieces in the format in which they will appear when printed, mainly for authenticity. When you put these pieces into a portfolio case, however, they cannot always be shown in their actual format. For example, if you produced a life-sized comprehensive of boxes for a line of computers, its outside package would most likely be too large for the portfolio case. You can, of course, remove the product and inner packaging and present the piece flat, but that solution would most likely detract from the effect achieved by the graphics.

The best solution for presenting large three-dimensional pieces is to record them as slides or transparencies. If many or all of your pieces are oversized, your entire presentation can be shown as slides or transparencies.

CHOOSING A PRESENTATION VEHICLE

A portfolio presentation vehicle is the actual case, book, or box that holds and protects all of your work for transporting and presenting it. The vehicle you choose should be determined by the needs of your work. If your work consists of artwork or transparencies mounted on board, a

Figure 6.9. Sketches that are rendered on thin transparent tracing paper are often matted *(left)* or mounted *(right).*

Figure 6.10. Thumbnail and early layout sketches can be photocopied and 3-hole punched for insertion in 3-ring binders.

rigid presentation box case is the best. This case will also accommodate items such as a three-ring binder holding 35-millimeter slides or sketches, or actual three-dimensional pieces.

The four most popular portfolio presentation cases used by graphic designers are multi-ring presentation books, multi-ring presentation cases, and two types of rigid presentation box cases (Fig. 6.11).

Multi-ring Presentation Books

Multi-ring presentation books are lightweight, portable books used to present layouts, photographs, comprehensives, and two-dimensional printed pieces. The cases are available in real or simulated leather finishes and contain clear acetate sleeves and range in size from 8½ by 11 to 18 by 24 inches. Pieces can be inserted vertically or horizontally, depending on their sizes and configurations. Multi-ring books enable you to make a neat, organized, uniform presentation.

Multi-ring Presentation Cases

Multi-ring presentation cases are similar to multi-ring books, except that they have a carrying handle and a zipper that seals the entire case. They also generally have pockets for carrying transparencies, slides, and bulky pieces.

Zippered cases also exist that do not have pockets or ring binders. These are suitable for storing and carrying pieces but not for presentations.

Rigid Presentation Cases

A rigid presentation case is a deep attaché-style box that comes with or

Figure 6.11. *Top to bottom:* multi-ring presentation book, multi-ring presentation case, rigid presentation box (wood and cardboard frame), and rigid presentation box (all-wood frame).

without inside compartments. There are two types: One type features an all-wood frame; the other is made of wood and cardboard and is therefore lighter.

Advantages and Disadvantages

Each portfolio has advantages and disadvantages, which should be considered before you purchase one. A main advantage of the multi-ring presentation book and case is that because they contain vinyl prepunched pages that are fixed in a binder, the pieces will always be viewed in the same sequence. They are also portable and lightweight. A disadvantage is that the acetate sleeves often create glare and may actually darken or alter the colors of the artwork. Also, the books cannot be used for showing three-dimensional pieces, such as packaging comprehensives and printed packages.

The main advantages of both types of rigid presentation case are their extra protection and their ability to accommodate a wide variety of mounted two-dimensional and unmounted three-dimensional pieces. Another feature of these cases is that individual pieces can be handled and viewed in the round. Also, because the pieces are mounted on individual boards, they can be taken out and inserted to suit the particular needs of a client or prospective employer. This flexibility lets you custom-tailor your portfolio quickly without the need for two or three separate portfolio cases.

The biggest disadvantage of the rigid type of case is its weight, especially when combined with the additional weight of mat boards, but for beginning designers, the rigid box type is a good all-around case. It can easily hold beginning portfolios that consist of comprehensives and perhaps some printed pieces. For more experienced graphic designers, who have mostly printed pieces, a lighter-weight, more compact portfolio presentation case, such as a multi-ring presentation book or case might be more appropriate (Fig. 6.12).

Figure 6.12. Before purchasing a portfolio case, assess the nature of work that it will contain. Generally, a multi-ring presentation case *(left)* is best for flat, printed work, while a rigid box *(right)* is best for comprehensives and unprinted two- and three-dimensional pieces that are matted and/or mounted on boards.

chapter 7

Demonstrations

The three assignments in this chapter are demonstrations of different types of graphic design problems, providing a cross section of the work of designers. They include a mass-market-oriented package, an annual report, and a magazine advertisement.

Each of the demonstrations uses equipment, tools, materials, and techniques described in Chapters 2 through 6, but because of their varied nature, each has its own approach while adhering to a framework that helped the designers to develop the ideas in an efficient, comprehensive manner.

Each demonstration begins with a statement of the problem or assignment and any design directions or restrictions, followed by the steps and procedures needed to solve the problem.

DEMONSTRATION ONE: PACKAGE DESIGN ASSIGNMENT

The problem to be solved by this particular designer was to redesign the Listerine antiseptic bottle (Fig. 7.1) and outside wrapper using either a close-in (or related to the package's original look) traditional approach or an up-to-date or contemporary approach, to result in a totally new appearance for the product. This problem was an actual competition sponsored by the New York–based Schechter Group, a leading design organization that specializes in package designs for major consumer brands. The assignment was hypothetical, and the Warner Lambert Company was not involved with the competition.

Step One: Research and Analysis

After reviewing the specifications and parameters provided by the Schechter Group, the designer visited a local drugstore in order to become familiar with other products in the same category. Listerine's memorable yellow and army green color combination seemed to set it apart from competitors. The designer therefore felt that an image that retained Listerine's distinctive look but upgraded and modernized its appearance could be a solution.

Next, the designer went to the New York Public Library to do research on antiseptics and oral hygiene. The designer found a book containing an old photograph of Joseph Lord Lister (the surgeon and scientist who revolutionized surgery with his antiseptic practices), and borrowed it along with material from the library's picture file collection (Fig. 7.2). After a thorough examination of the specifications for the assignment and the gathered material, the designer was ready to do thumbnail sketches.

Step Two: Thumbnails

In the thumbnail stage, the designer explored many design directions and formats for the Listerine package itself and the graphics that would be applied to it (Fig. 7.3). Virtually anything that came to mind was recorded quickly and loosely to provide a wide range of ideas from which rough layouts could be developed.

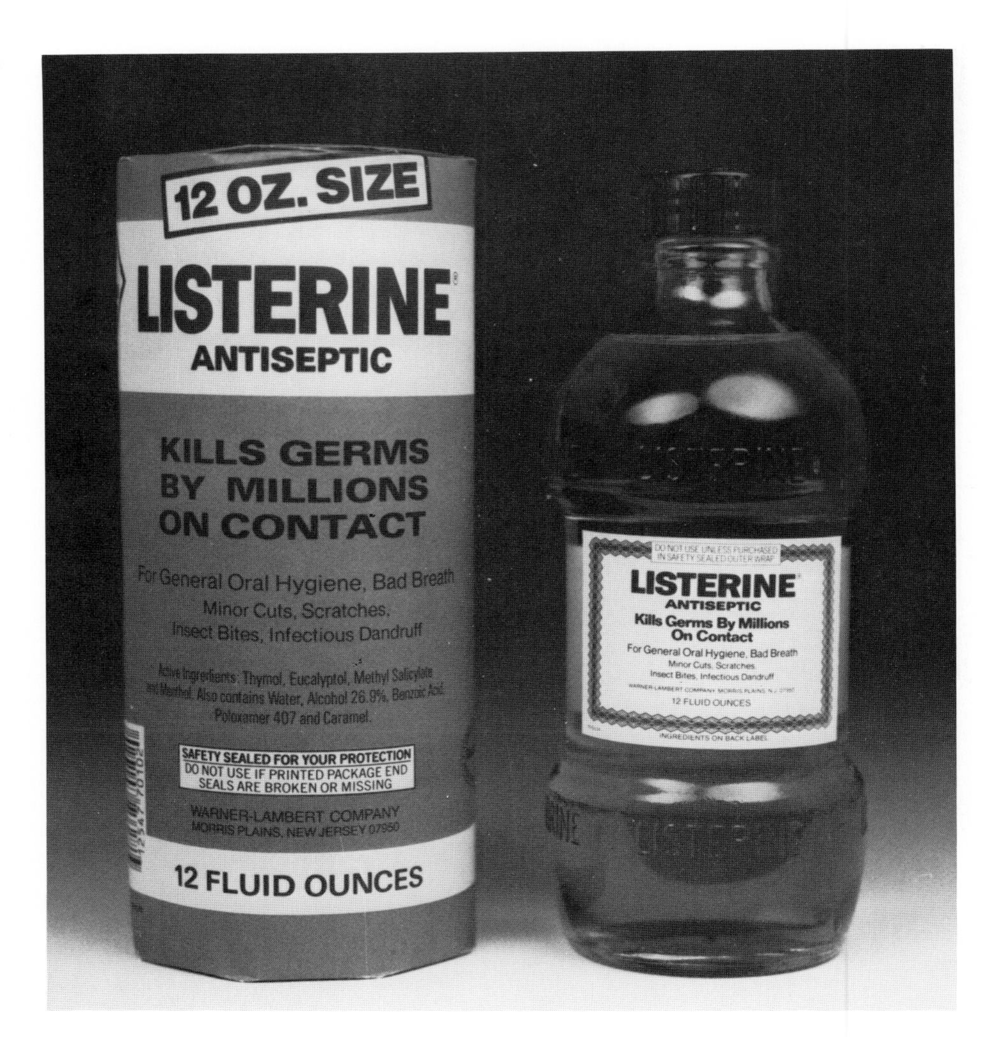

Figure 7.1. The existing Listerine antiseptic bottle and its outside wrapper.

Figure 7.2. Visual research materials gathered from the New York Public Library's picture collection and other sources.

Figure 7.3. Thumbnail sketches should be recorded quickly, in a loose manner, in order to accumulate a large volume of ideas.

Figure 7.4. Thumbnails should be enlarged to reproduction size. In this case, the thumbnails were enlarged by visual estimation.

Step Three: Rough Layouts

After completing a dozen or more thumbnail sketches, the designer chose the ones to be enlarged into rough layouts (Fig. 7.4). Those thumbnail designs edited for later refinement mainly included designs that looked similar to the original design yet more contemporary. This decision was made by the designer after reviewing the assignment criteria again and after showing some thumbnails to the instructor (who served as the client in this case).

Since a rough layout's main function is to show spatial relationships among the design elements in a light, loose manner, tools and media appropriate for such spontaneity were used: The roughs were done on tracing paper, which is resilient enough to stand up to spirit- and water-based markers without bleed-through; the text and display type were drawn freehand to allow for fast and easy experimentation (Fig. 7.5).

As each rough layout was completed, it was placed under a fresh sheet of tracing paper, new dimensions were traced in pencil, and further refinements were made. After many layouts were done, the most promising ones were edited, and the rest were filed away for reference.

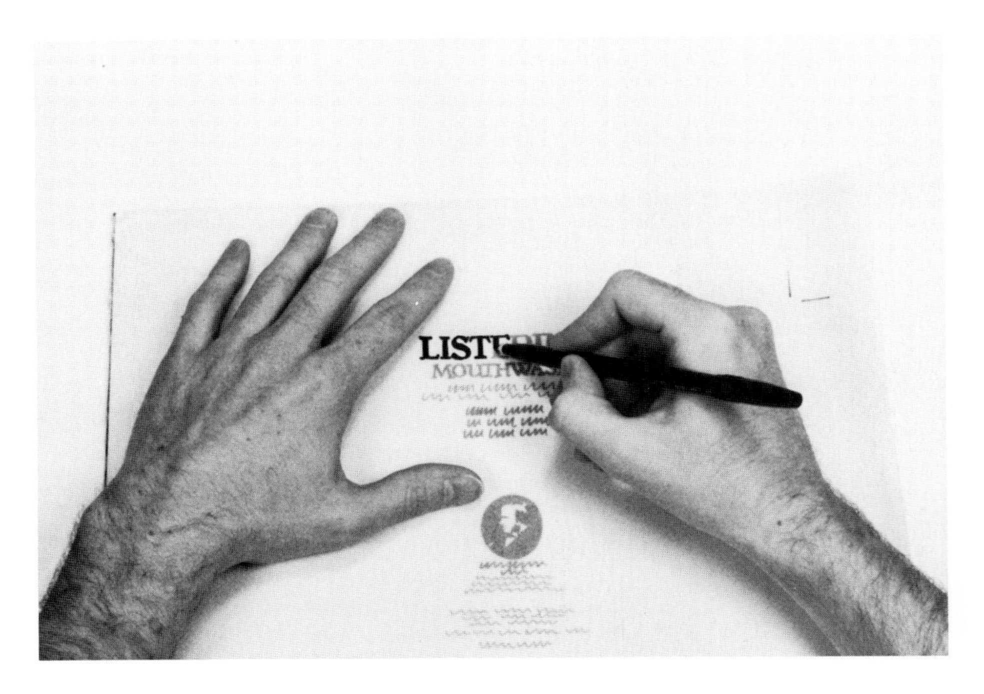

Figure 7.5. All text and display (or headline) type are drawn freehand without the use of drawing or straightedge guides. As each rough layout is completed, it is placed underneath a fresh sheet, new dimensions are traced, and further refinements are made.

Figure 7.6. The best rough layouts are refined by placing previously completed sketches under a clean sheet of tracing paper, tracing off new dimensions in pencil, and making any adjustments.

Step Four: Finished Layouts

The promising rough layouts were evaluated and refined through trial and error, reducing the number of ideas further. As in the rough layout stage, refinements were made by placing completed sketches under a clean sheet of tracing paper, tracing the new dimensions in pencil, and then making adjustments (Fig. 7.6). Finally, one rough layout was chosen. It contained a silhouette portrait of Lister and a spot label, a label that does not wrap completely around the container but is affixed only to the front side of the product with adhesive. Beer bottles generally use spot labels for identification.

Once the position of the design elements was established in black-and-white, using a T square and triangle to align elements vertically and horizontally (Fig. 7.7), color was introduced (Fig. 7.8). Yellow, green, and black—the colors of the existing Listerine wrapper—were indicated with Design Art Markers on the finished layout. Green and yellow bands were used for the background elements, and black was used for the foreground.

The background, which was a large mass of color, was rendered with wide-nib spirit-based markers using the T square as a horizontal guide. 3M Post-It note pad sheets were adhered to each side of the layout to prevent marker build-up at the beginning and end of the strokes and so that the edge would be cleanly made. After the background colors were applied, the black-and-white finished layout was slipped underneath the colored layout. A T square and triangle were used to draw horizontal and vertical guidelines for display and text type indication. Next, the foreground text elements were indicated with a fine-tipped black Marvy Marker.

Figure 7.7. When rendering a finished layout, use a T square and triangle to align all elements vertically and horizontally.

Figure 7.8. Once the position of the design elements has been established in the remaining layouts, color can then be experimented with further. When applying a large mass of color with markers, 3M Post-it notes can be adhered to the layout to prevent marker ink build-up at the beginning and end of each stroke.

Figure 7.9. If the comprehensive layout will be rendered solely in color markers, square up a 100% Rag Layout Marker Pad onto the drawing table top. Slip a finished layout under the top sheet of the pad, and carefully render the entire design using a T square and triangle as straight-edge guides. Repeat this process for all remaining finished layouts.

Figure 7.10. Design Art Markers are used for rendering large background areas.

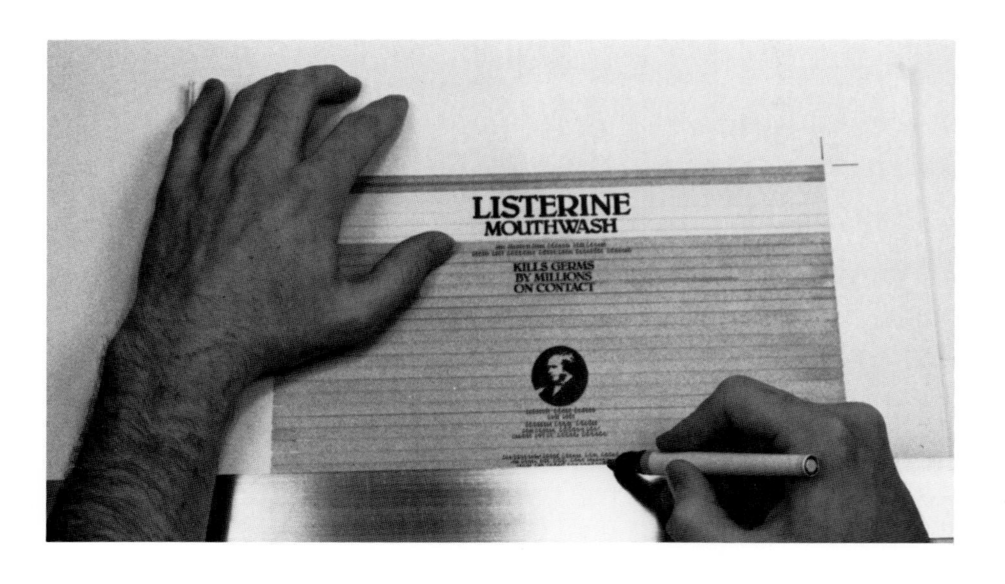

Figure 7.11. A 3 × 0 technical pen is used to indicate the text type, to outline display type and the silhouette shape. Fill in these outlined areas with a fine-tipped black Marvy Marker.

Figure 7.12. Trim the comprehensive layout and wrap around the actual three-dimensional shape on which the graphics would eventually be applied.

Figure 7.13. View a group of comprehensive layouts together for comparison, or actually place them on a store shelf to evaluate their effectiveness. If there are any further changes to be made of the design itself, they should be executed at this point.

Step Five: Comprehensive Layout

Using the finished layout as an underlying guide, a comprehensive layout was rendered in color markers on a 100% Rag Layout Marker Pad (Bienfang #360) (Fig. 7.9), which does not bleed and retains the color, brilliance, and sharp edges. A light box was used to study the fine details on the underlying finished layout.

When rendering the comprehensive, a T square and triangle were used to ensure straight vertical and horizontal lines, and an oval template was used to draw the shape for the silhouette. Broad-nib Design Art Markers were used to render the background colors (Fig. 7.10), and a 3 × 0 technical pen was used to indicate the text type and to outline the large display type and the silhouette (Fig. 7.11). The outlined portions were then filled in with a fine-tipped black Marvy Marker.

After the comprehensive was rendered, it was cut to size and wrapped around a hand-crafted white cylinder dummy so that it could be viewed in the round (Fig. 7.12). To compare it accurately with competitors, it was then placed on a drugstore shelf (Fig. 7.13).

Step Six: Advanced Comprehensive

Next, the materials were prepared to render the interim mechanical for the advanced comprehensive. Artwork and typography were refined and tightly rendered as if the piece were to be printed. Type was burnished from Letraset dry-transfer lettering sheets onto illustration board with a fine-tip burnisher (Fig. 7.14). The silhouette of Joseph Lister was going to be dropped out, so the ellipse had to be inked on multipurpose drafting vellum using a 3 × 0 technical pen and an ellipse template (Fig. 7.15). Then a photostat was made from the original photograph of Lister to create a "posterized" high-contrast effect without middle gray tones. The image was then cut out to fit within the elliptical shape. The edges of the cut photostat were blackened with a broad-nibbed permanent marker and adhered to the inked ellipse shape with spray adhesive. All the elements were prepared separately and then photostatted to the sizes indicated on the comprehensive.

The interim mechanical board was then prepared at reproduction size to accommodate the outside wrapper and spot label side by side. Guidelines were drawn in pencil and crop marks were inked using a 3 × 0 technical pen, a T square, and a triangle (Fig. 7.16). The elements were photostatted and cut out using a T square, triangle, and art knife and adhered to the interim mechanical board using one-coat rubber cement (Fig. 7.17).

Next, an I.N.T. transfer sheet had to be made. A direct positive photostat was made of the spot label (Fig. 7.18). Then a negative film was made of the outside wrapper. The photostat for the spot label was trimmed, the back was coated with rubber cement, and it was adhered to the bottle, which was coated with rubber cement (only in the label area) to provide a permanent board. A wax paper slip

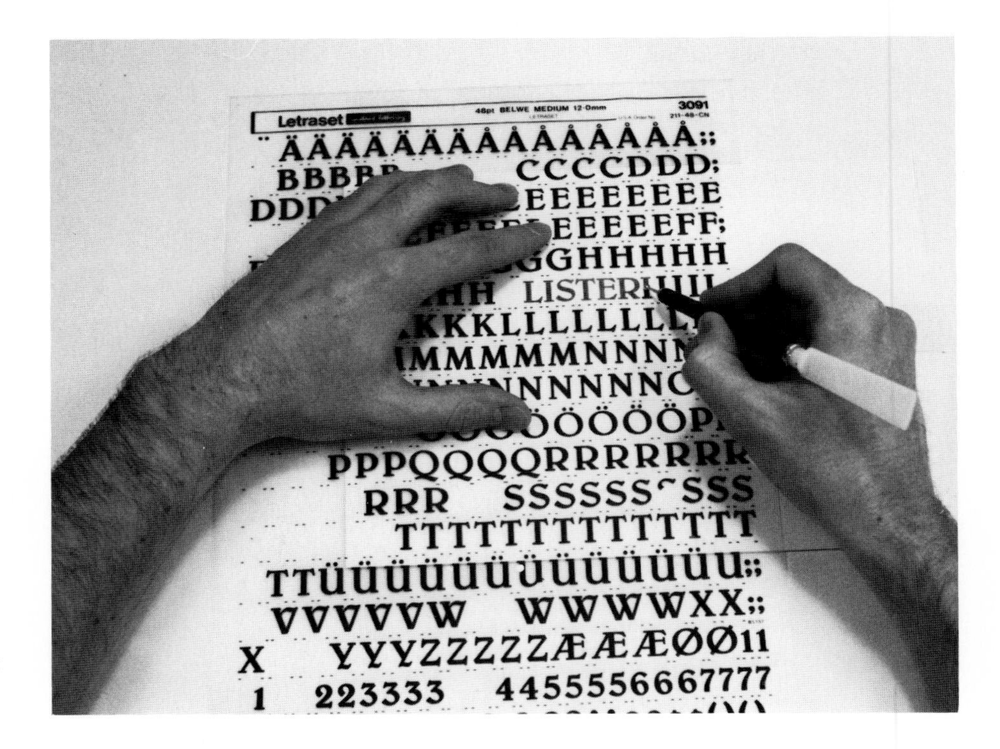

Figure 7.14. If set type is not available or affordable, transfer lettering such as Letraset can be burnished from dry-transfer sheets onto a separate piece of illustration board.

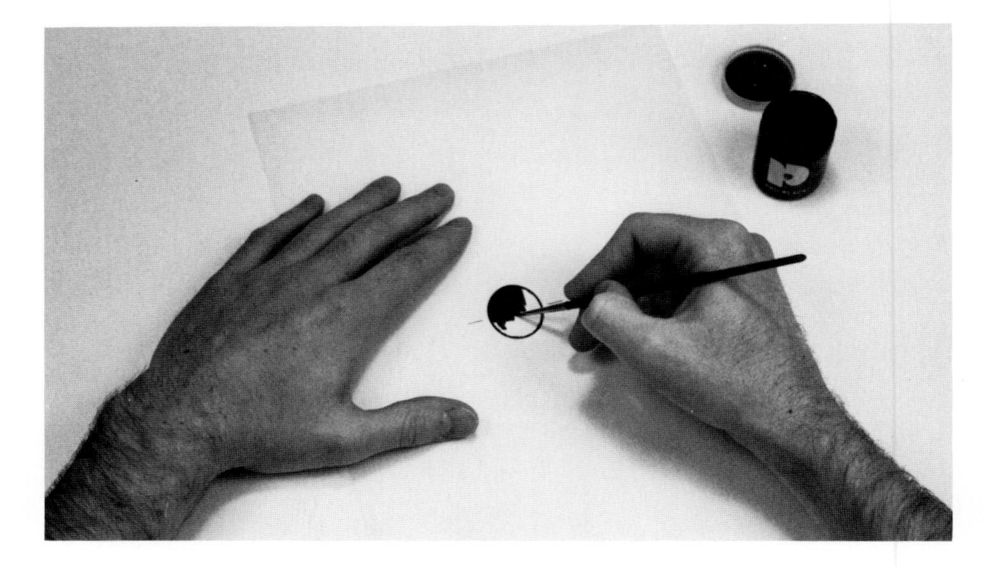

Figure 7.15. Any artwork that has to be created should first be tightly drawn on tissue layout paper and then inked on a piece of multipurpose drafting vellum before photostatting. Here the outline of an oval was drawn using an ellipse template and 3 × 0 technical pen. The shape was then filled in using a brush loaded with India ink.

Figure 7.16. Choose a board that will accommodate the size of the artwork plus a minimum of a 1 inch border all around. This allows for easier handling and better protection of the artwork. Indicate all guidelines with a graphite or blue lead pencil, and then draw in crop marks with a 3 × 0 technical pen.

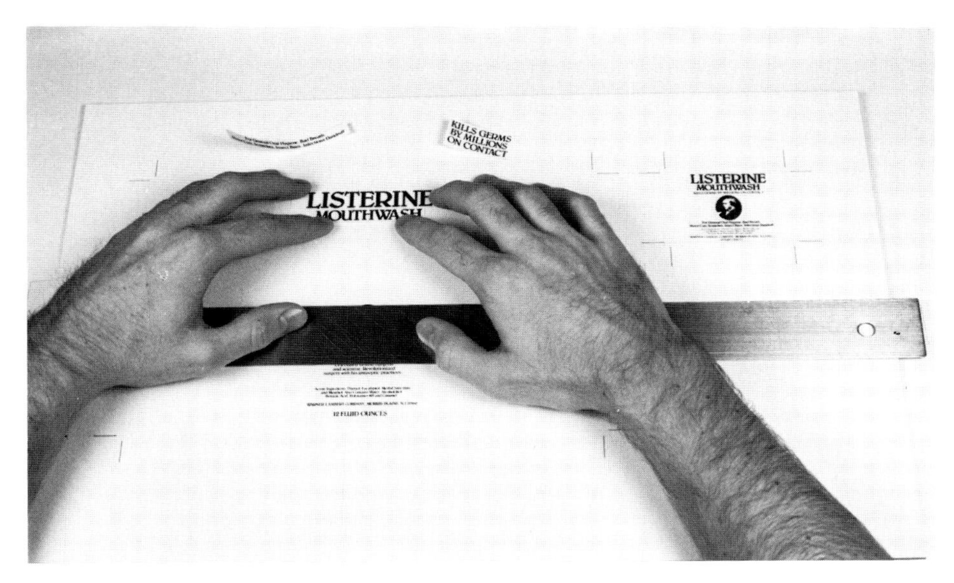

Figure 7.17. Cut out and adhere all of the photostatted pieces onto the interim mechanical board in the position indicated on the comprehensive layout.

Figure 7.18. Since the advanced comprehensive for the spot label will be rendered as a photostat, its interim mechanical can simply be shot as a direct positive photostat. The outside wrapper mechanical, however, has to be shot as a negative film in order to make an I.N.T.

sheet was wrapped tightly around the midsection of the bottle and held in place with white tape to ensure a tight permanent bond while the cement dried. The negative film was contacted, exposed with a sheet of black I.N.T., developed, and allowed to dry.

While the I.N.T. was drying, a blank dummy was made of the outside wrapper by reinforcing a corrugated cardboard cylinder with three mat board disks and placed inside for support. A sheet of heavyweight bond paper was then adhered around the corrugated cardboard to hold the pieces in place and provide a clean surface on which to mount the comprehensive.

The outside wrapper of the existing Listerine package was measured, and its outside dimensions were drawn in pencil on a sheet of green, uncoated Pantone paper. Next, two yellow bands were cut out of uncoated Pantone paper and adhered to the green base sheet in the position indicated on the comprehensive layout.

Once the I.N.T. had dried, it was placed over the background papers and hinged along its top edge so that it would not slip during burnishing. (A slip sheet is helpful at this stage because it allows easy positioning but does not let the I.N.T. accidently adhere to any underlying surface.) The transfers were then burnished down section by section from top to bottom; the clear film backing was constantly lifted as each portion was transferred (Fig. 7.19). After the entire I.N.T. was transferred, a wax paper slip sheet was placed over the comprehensive and additional pressure was gently applied with the plastic end of the burnisher so that the transferred I.N.T. would not be damaged.

The extra space at the top and bottom of the advanced comprehensive was used to score and cut tabs to secure the wrapper to the blank dummy. The top was then flipped over so the top and bottom edges could be scored using the plastic end of the Zipatone burnisher; this ensured that the multiple folds would look neat and uniform when the wrapper was assembled (Fig. 7.20). Double-sided tape was applied to the one-half inch overlap on the top and bottom. Cuts were made through the tape and wrapper down to the scored line to maintain the round cylindrical shape during wrapping. Two strips of double-sided tape were then attached to the left and right sides of the wrapper's underside—one to grip the wrapper to itself. The taped wax paper slip sheet and any glue residue around the spot label were removed from the Listerine bottle (Fig. 7.21).

The wrapper was then adhered to the dummy in the proper position, leaving the top and bottom open (Fig. 7.22). The bottom tabs were folded down one over the other in a clockwise direction until they were firmly adhered to the dummy. The same process was repeated for the top. Two paper disks were then cut out of matching green Pantone paper with a beam compass cutter (see page 30) and were adhered with spray adhesive to the top and bottom panels to cover the overlapping flaps of paper. Finally, both advanced comprehensives—for the package and the design—were photographed and inserted into a portfolio (Fig. 7.23). Figure 7.24 shows the new (left) and existing (right) Listerine packages.

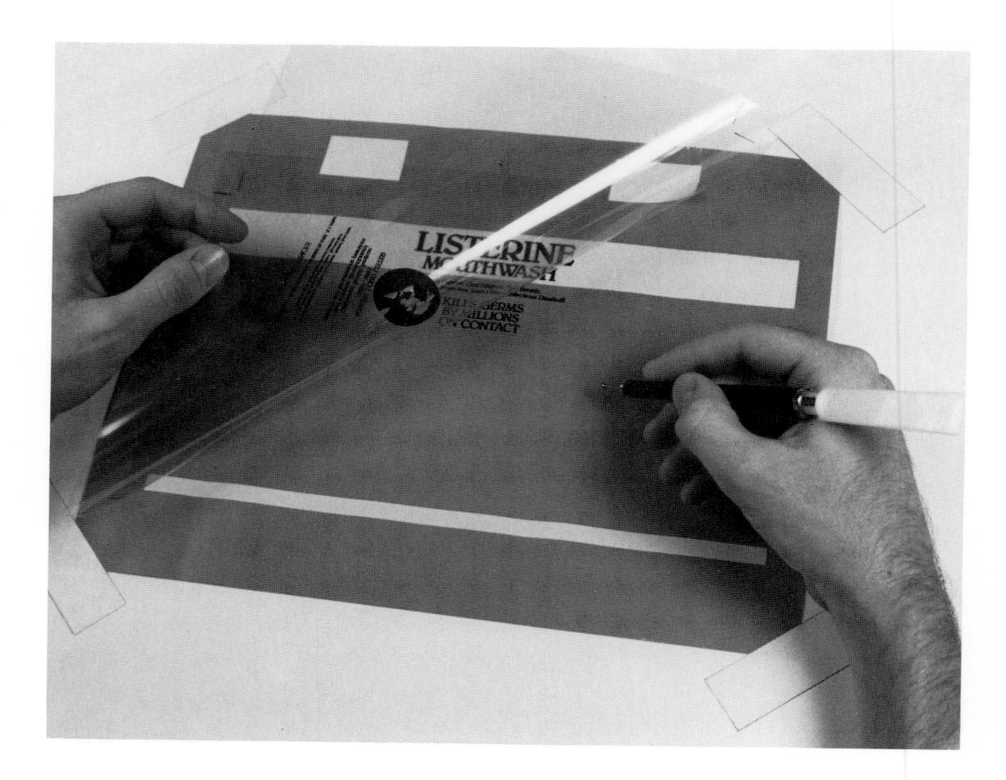

Figure 7.19. Burnish down the I.N.T. section by section from top to bottom.

Figure 7.20. Trim and score the comprehensive.

Figure 7.21. Remove the wax paper slip sheet and any glue residue from around the spot label.

Figure 7.22. Adhere the comprehensive securely onto the dummy with double-sided tape or a similar strong adhesive.

Figure 7.23. Both comprehensives can now be presented or photographed for insertion into a portfolio.

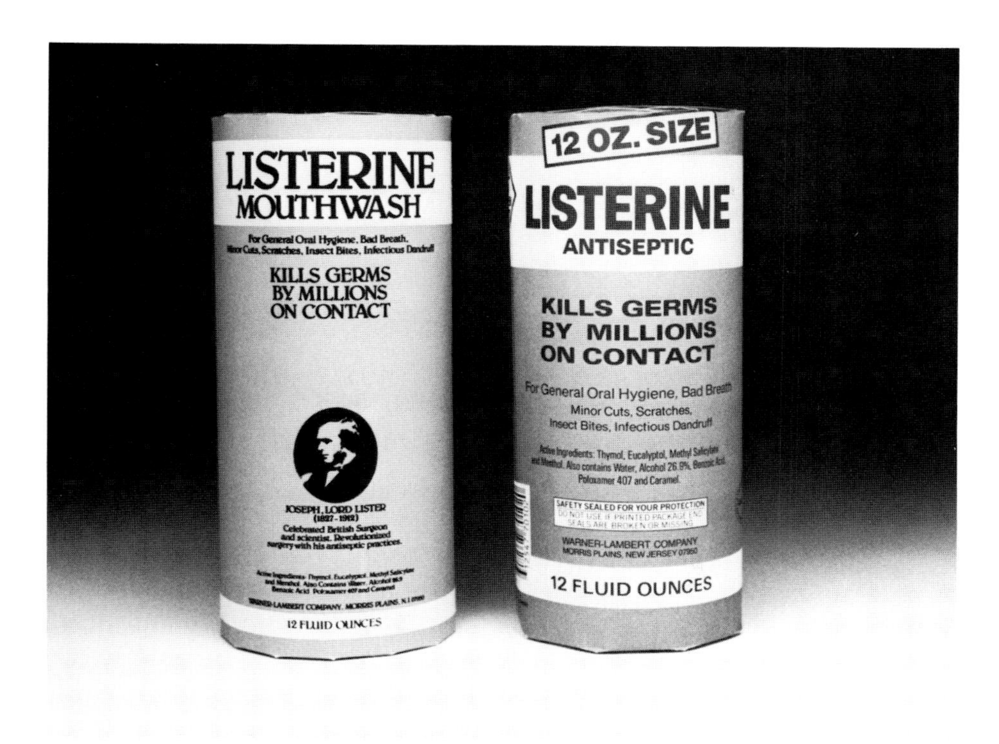

Figure 7.24. The new *(left)* and existing *(right)* Listerine packages.

Figure 7.25. A helpful step before approaching the design of an annual report is to become familiar with the range of current annual reports in a similar category. A good source of reference and inspiration is the numerous graphic design periodicals and annuals.

DEMONSTRATION TWO: ANNUAL REPORT ASSIGNMENT

This problem was a school assignment to design a new corporate identity for a company chosen by the student, with an annual report as the premier promotional piece. A single theme had to be developed for the report in a cover design and four typical sample spreads. The spreads were to include a two-year summary, a president's letter, a research and development piece (or a public relations-oriented spread), and a financial spread. The format and content were left up to the designer, as long as the necessary sections of the annual report were included.

The designer decided to develop a new logotype for an existing company, the Carvel Corporation, launching the beginning of a new corporate identity. The graphic standards established for the company were then to be used to design the annual report.

Step One: Research and Analysis

In order to become familiar with annual reports, the designer began by examining recent graphic design books and annuals for layout, format, and content. Copies of annual reports were borrowed or ordered from various companies (Fig. 7.25). Reviewing design annuals and annual reports at this early stage provided inspiration and reference and started the flow of ideas.

Additional research was done by examining the New York Public Library's picture file on the subject of ice cream (Fig. 7.26). The designer was able to find and borrow interesting line drawings of ice cream-making equipment and old photographs of street vendors and roadside stands. During this early stage, the designer considered various themes for the annual report, including new product development and services, customer testimonials (such as statements given after sampling Carvel's ice cream products), and growth as an emerging twenty-first-century company. The third theme was chosen because it promoted the idea of a contemporary, changing company, which was reflected in the new corporate identity.

Figure 7.26. Visual research materials gathered from the New York Public Library's picture collection.

Step Two: Thumbnails

While reviewing materials during the research and analysis stage, the designer made notations and scribbles on 3M Post-It note pad sheets. These individual sheets were then compiled, edited, and incorporated into thumbnail sketches on an 11-by-14-inch tracing paper pad with a fine-tipped marker (Fig. 7.27a).

Different formats for the annual report were explored rapidly in sketch form. Square, rectangular, vertical, and horizontal formats were drawn freehand, with scribbles indicating type and pictorial elements. All the inside pages were sketched as spreads, which is how they would be presented. Unifying design elements such as rules, borders, colors, typography, and photography were explored to establish a rhythm and common theme. The cover designs were developed at the same time to ensure a unified look (Fig. 7.27b).

a

b

Figure 7.27. *a.* Different formats and concepts for the annual report should be explored rapidly in sketch form. *b.* By developing cover designs simultaneously with the inside pages, a more unified look will result.

Figure 7.28. Square up the tracing pad onto the top of the drawing table. Indicate the dimensions of the intended reproduction size of the piece (or pieces) that were chosen from the thumbnail stage. Because of the large pad size required to draw a double-page spread at actual size, these rough layouts are sketched at one-half the intended reproduction size.

Figure 7.29. Enlarge the preferred thumbnail sketch (or sketches) to these dimensions, making any design adjustments that are required in order to fit.

Step Three: Rough Layouts

The designer began by squaring up a tracing pad to a horizontal position to fit a double-page spread. Because a large pad was needed to draw the double-page spread, the rough layouts were sketched at one half the reproduction size (Fig. 7.28). Thumbnail sketches were enlarged using a Lucy and by visual estimation (Fig. 7.29). The most promising thumbnails were revised as rough layouts made to the correct dimensions: Since the reproduction size of a single page would be 11 inches square, the rough layouts were drawn as 5½-by-11-inch double-page spreads. A T square and triangle were used to indicate the exact half-size dimensions. The designer then experimented with spatial relationships among the design elements in a light, loose manner, using a grid underneath as an aid to organize and align typography, photography, and design elements (Fig. 7.30), taking into account the amount of text, and size and the number of illustrations. A four-column grid was chosen since it was flexible, allowing for the financial sections of the annual report. The exact dimensions of the grid would be fine-tuned later when it would be enlarged to reproduction size.

As each rough layout was completed, it was placed underneath a fresh sheet of paper, its dimensions traced and any further adjustments made. The best of these rough layouts were then edited to include only the strongest design concepts (Fig. 7.31).

From the sketches, four sample spreads and a cover design, which were reasonably resolved in terms of color, size, and position of design elements, were chosen. These roughs were then ready for enlargement to reproduction size as finished layouts.

a

b

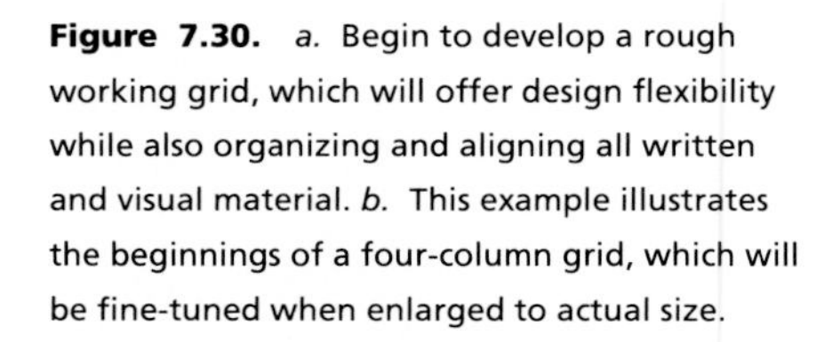

Figure 7.30. *a.* Begin to develop a rough working grid, which will offer design flexibility while also organizing and aligning all written and visual material. *b.* This example illustrates the beginnings of a four-column grid, which will be fine-tuned when enlarged to actual size.

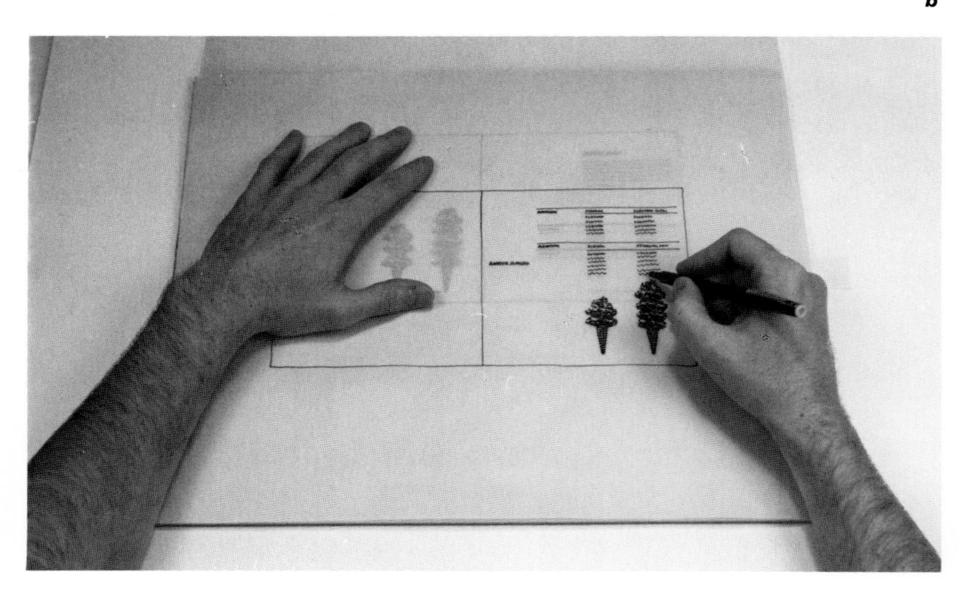

Figure 7.31. Refine design elements on the rough layout.

Figure 7.32. Final grid for a double-page spread.

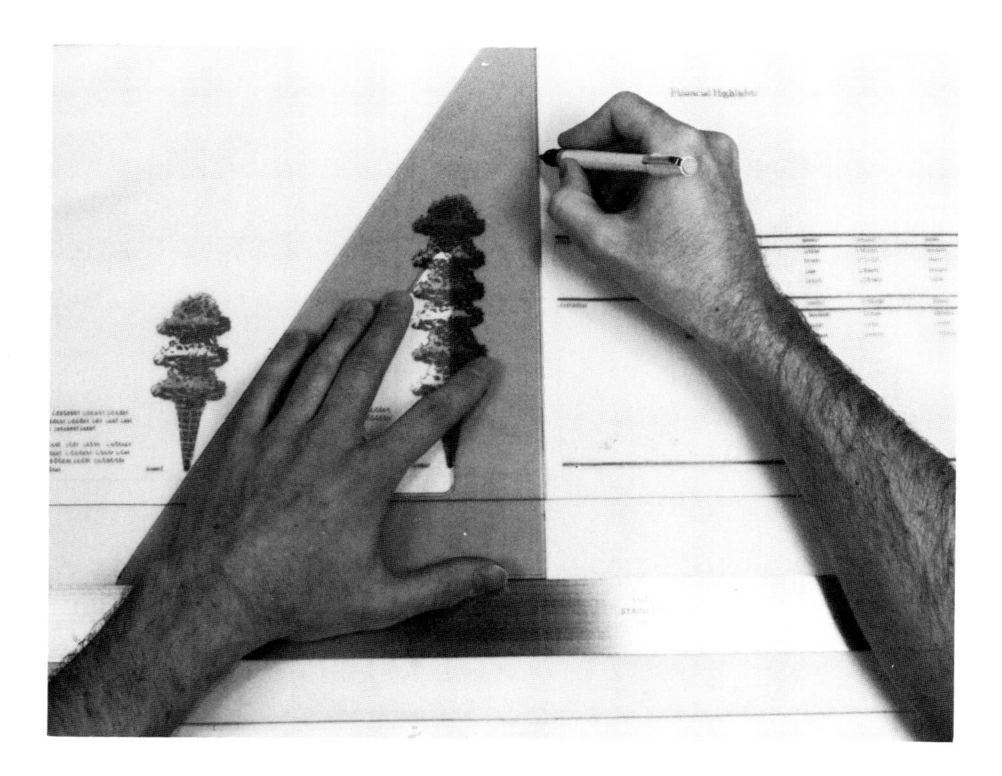

Figure 7.33. After all the rough layouts have been enlarged to actual size, refine each design individually by tearing off and placing previously completed sketches under a clean sheet of tracing paper, tracing off new dimensions in pencil, and making any design adjustments that need making.

Step Four: Finished Layouts

The grid, the cover design, and the four sample spreads were enlarged to reproduction size with a Lucy, which is more accurate than visual estimation and less expensive than photostatting. Another advantage of the Lucy is that size adjustments and design alterations can be made easily while enlarging the rough layouts.

Once the rough layouts were enlarged to reproduction size, the tracing paper pad was squared up onto the top of the drawing table, and the dimensions of a final grid for a double-page spread were drawn (Fig. 7.32). All of the design elements of the enlarged layout were then traced in position within these dimensions. Next, refinements were made by placing the sketches under a clean sheet of tracing paper, tracing the new dimensions in pencil, and making further changes (Fig. 7.33). This process was repeated several times for the cover and each spread until they worked individually and as a group.

Line art was rendered in fine- or broad-nib colored markers in a loose manner with positions accurately indicated (Fig. 7.34). Photographs were indicated as tones rather than outlined shapes to simulate their weight when reproduced. These sketches were done on separate pieces of paper, which were later integrated into the layouts. The text type was indicated as loops; baseline guides were drawn using a marker and T square and a type gauge to ensure accurate baseline to baseline measurements (Fig. 7.35).

All the design elements on the cover and each of the spreads were compatible in their level of finish. At this time, the designer decided on an upscale typographic treatment, using the new logotype on the cover, which would also heighten the logo's visibility (Fig. 7.36). The logo would also be hot-stamped.

Furthermore, to tie together the look of the cover and that of the inside spreads, the designer experimented with the graphic use of color, scales, dots, bands, and bars as repeating elements.

Next, the background shape was rendered using the broad nib of a red Prismacolor Art Marker and a T square as a straightedge guide. 3M Post-It note pad sheets were attached to the layout so that smooth color application could be obtained without ink build-up at the beginning and end of the marker strokes. Using an extra-fine-nib metallic silver Pilot permanent marker, the calligraphic logotype was rendered over the red background. Type was then indicated with a 3 × 0 technical pen using a T square as a baseline guide. To get a true feeling of the typeface, each letter was traced individually, in Times Roman, from the text type section of a photolettering catalog.

Figure 7.34 *(top).* Render all photography and line art in marker or colored pencil in a loose manner, but indicate their positions within the layout accurately.

Figure 7.35 *(middle).* All text type can be indicated with a T square as a baseline guide. Use a type gauge to ensure accurate baseline-to-baseline measurements.

Figure 7.36 *(bottom).* The cover should be rendered to the same degree of finish as the inside spreads.

Step Five: Comprehensive Layout

Because of time restrictions and accessibility of the equipment needed to reproduce an advanced comprehensive, the comprehensive layout stage was omitted for this assignment.

Step Six: Advanced Comprehensive

After the final layouts for the cover and page spreads were finished, the best ones were further edited to include the strongest possible design elements. The most successful layout for each of the four spreads and cover were then chosen to be made into interim mechanicals for advanced comprehensives.

The typography was assembled and rendered at or close to reproduction quality. The rules were inked with a 3 × 0 technical pen on multipurpose drafting vellum, and the type was burnished in place under the rules (Fig. 7.37). Some color prints were enlarged from transparencies or negatives in the sizes indicated in the finished layouts. Others were clipped out of magazines or other sources and used at actual size. All line art for the advanced comprehensive was photostatted to the size indicated on the finished layouts. Display type was transferred onto bond paper using Letraset dry-transfer lettering and was photostatted to size; text type was transferred onto another piece of bond paper using dry-transfer body copy. A ragged right line break was inserted on the right edge of each column in order to simulate flush left, ragged right type, and paragraphs were indented.

The front cover design included a red rectangle bleeding off a square cream-colored background, with a white Carvel logo dropped out of the red area. There was black type positioned under the red rectangle. Thin, glossy coated paper was chosen for the background and flint paper (thin, lightweight glossy paper available in several colors) for the red rectangle.

A rub-down transfer of the logo was ordered to match the background color. Since all the type was black, Letraset transfer lettering would be burnished directly onto the background paper. To get the rub-down transfer of the logo, the designer prepared the artwork by photostatting the logotype, which was created at an earlier date, and sending it, along with a Pantone color swatch that matched the background color, to the supplier.

Next, the interim mechanicals for the inside spreads were prepared. The boards were trimmed to a manageable size, and outside guidelines were drawn with a sharp nonreproduction blue lead rod in a lead holder. Next, the inside guidelines were drawn, following the four-column grid configuration decided on in the finished layout stage. On each one-page grid, a column measured 14 picas wide, with 2 picas between each column and 2 picas as the left and right margins. A Schaedler Precision Rule was used to mark the measurements in picas. Crop marks were added in ink using a 3 × 0 technical pen so that there would be reference points for cutting (Fig. 7.38). Clear acetate was hinged over the base board of the interim mechanicals of the cover and one spread, which needed an overlay.

After the guidelines were drawn on the boards, the separate photostatted pieces were pasted up in the positions indicated on the finished layout (Fig. 7.39). Spray adhesive was used for tiny, hard-to-handle elements such as page numbers, and one-coat cement was used for large elements such as galleys of text type. Any pieces of line art that were going to appear on the comprehensive in color as 3M I.N.T.'s were placed on separate overlays. The overlays were hinged along the top edge of the interim mechanical base board using white tape. Then registration marks were placed on the base board and the acetate overlays (Fig. 7.40).

Since black-and-white photostats were going to be used as a base to simulate white printing paper, only type and line art that would be indicated in black-and-white on the advanced comprehensive were adhered to the board. Then a same-sized paper positive photostat of each interim mechanical base board was made. In this particular case, these photostats were used as a base on which other advanced comprehensive techniques would be applied. Next, colored photographs, which were ordered to the sizes indicated on the finished layouts, were trimmed using a T square and triangle as straightedge guides; silhouetted photographs were cut freehand with an art knife and using a #11 blade. The photographs were then cemented into place on the photostats with one-coat rubber cement (Fig. 7.41).

After each board was complete, it was photostatted, and a positive film, which would be used as a guide to position design elements or produce advanced comprehensive media, was made. The positive films, which contained the color line art on overlays, was sandwiched with a negative-orange Color Key to produce a negative. These negative films were then sandwiched with individual sheets of 3M I.N.T.'s to produce the desired colors.

To position the line art rendered as I.N.T.'s accurately, the positive films were hinged with white tape strips along the top edge of the board containing each advanced comprehensive. This made it possible to slide the I.N.T.'s under the film to the correct position before burnishing them down (Fig. 7.42). A wax paper slip sheet was used under the I.N.T.'s to prevent accidental image transfer. Once the I.N.T.'s were positioned correctly, the slip sheet was removed and the transfers were burnished in place.

After the four spreads were rendered as advanced comprehensives, they were trimmed to size using a T square, triangle, and art knife, following the crop marks as guidelines. They were then turned over, their

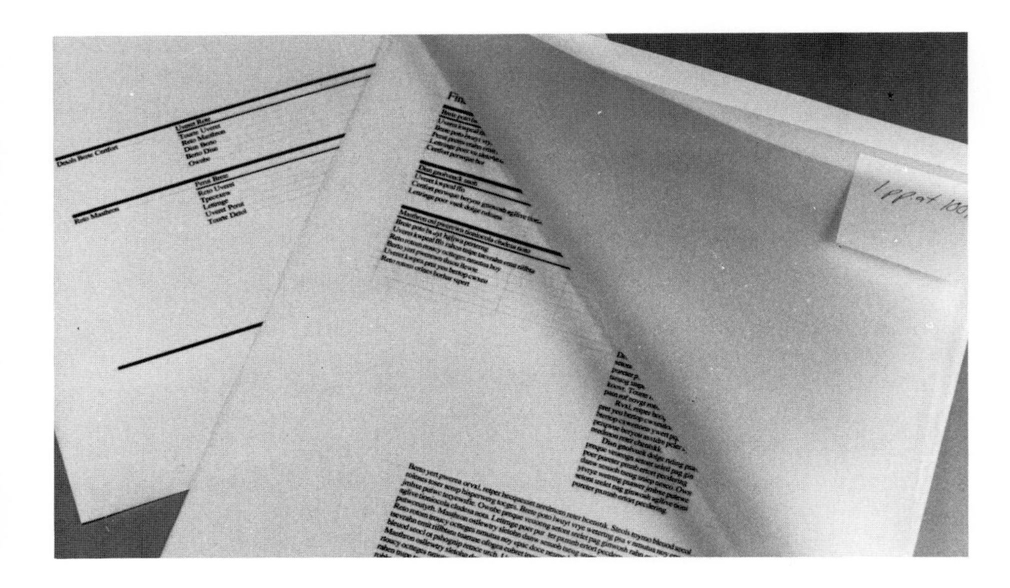

Figure 7.37. Type and rules should be rendered close to reproduction quality.

Figure 7.38. Using the underlying pencil markings as guidelines, draw crop marks with a 3 x 0 technical pen.

Figure 7.39. Cut out and adhere all of the photostatted pieces onto the interim mechanical boards in the position indicated on the finished layouts.

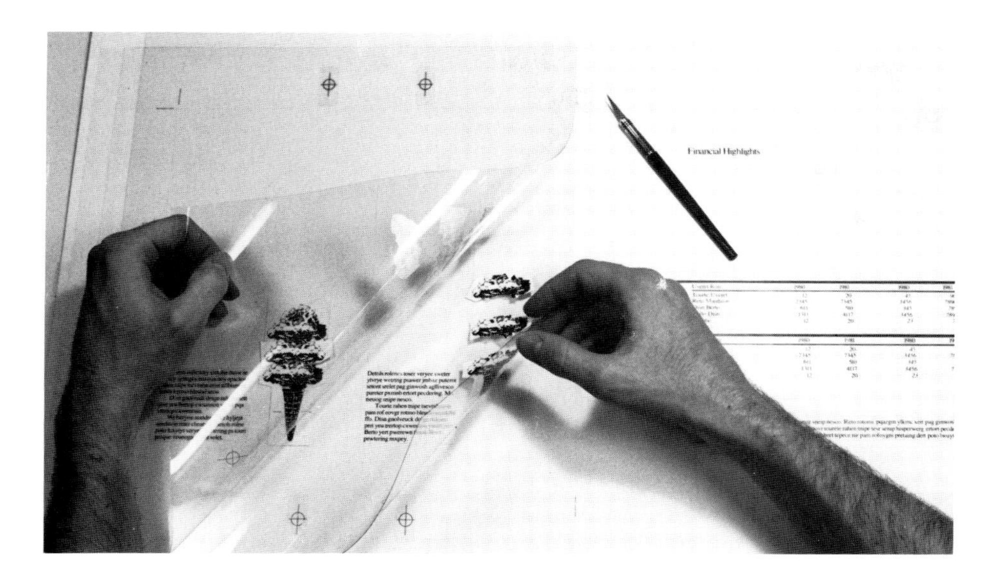

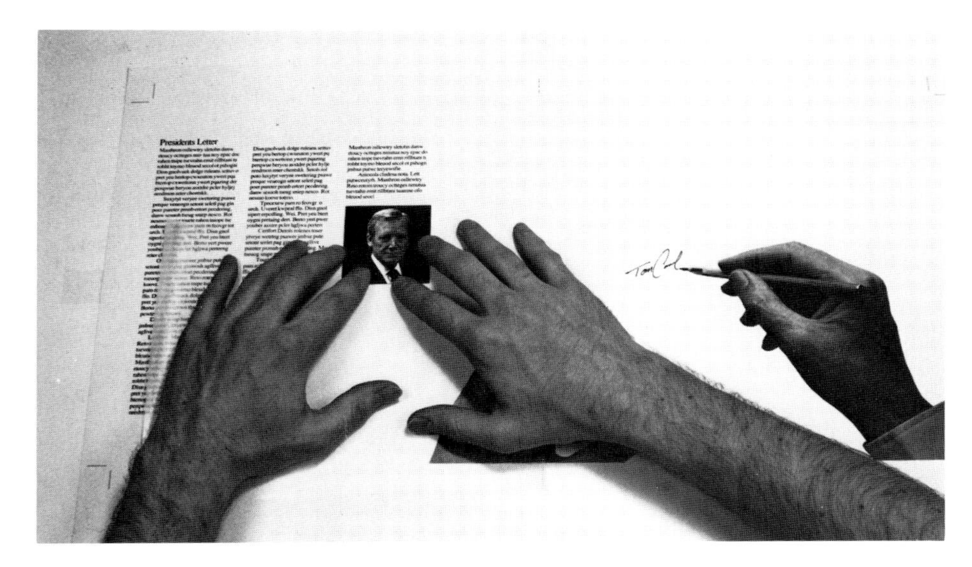

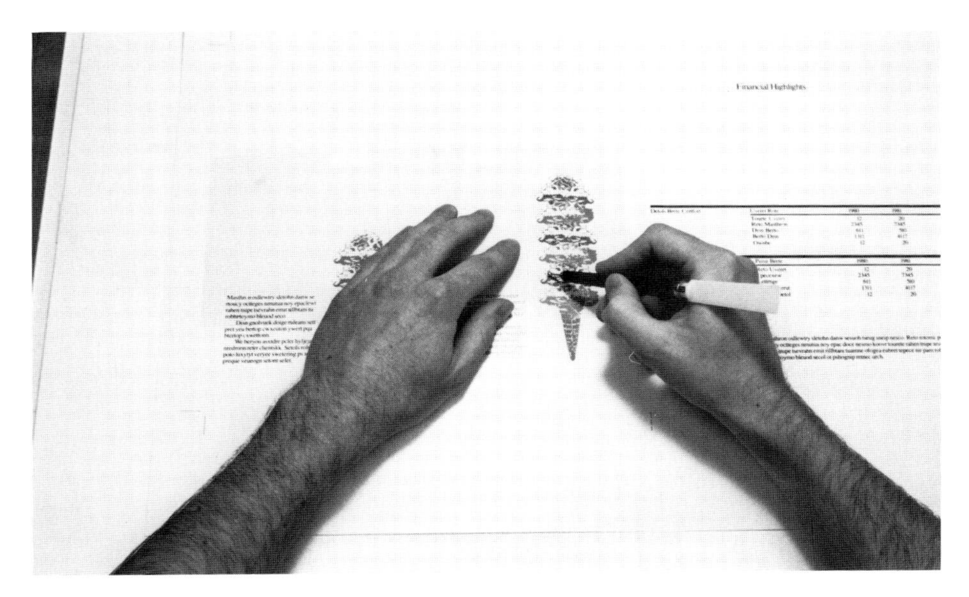

backs were coated with rubber cement, and they were adhered to a black mat board coated in one section with rubber cement, and were left to dry (Fig. 7.43). Excess rubber cement was then removed with a pickup.

In preparing the advanced comprehensive for the cover, thin cream-colored background paper was adhered to a sheet of heavyweight bond paper for support and as a base for cutting and burnishing. The red flint paper was cut slightly larger (by about one half inch all around) than what would be needed and was adhered to the background paper. A common cut was made through the flint paper and the background paper so that they butted together tightly. Any excess adhesive and paper were removed. The Letraset lettering was then burnished down on the background paper using a pencil baseline as a guide.

Next, the logotype was cut out and removed from its transfer sheet using an art knife. The positive photolith film (which was made through all of the overlays and the base board of the mechanical) was positioned on the background paper and hinged along the top edge with white tape. The logotype was then placed on a wax paper slip sheet (with the top half inch extending beyond the wax paper) and positioned using the photolith film as a guide (Fig. 7.44). It

Figure 7.40 *(top)*. Place line art that will appear in color as 3M I.N.T.'s on separate overlays.

Figure 7.41 *(middle)*. Trim the photographs to size and cement them into place on the photostats using rubber cement or a similar strong adhesive.

Figure 7.42 *(bottom)*. Hinge the positive film over each corresponding layout in order to position I.N.T.'s accurately before burnishing.

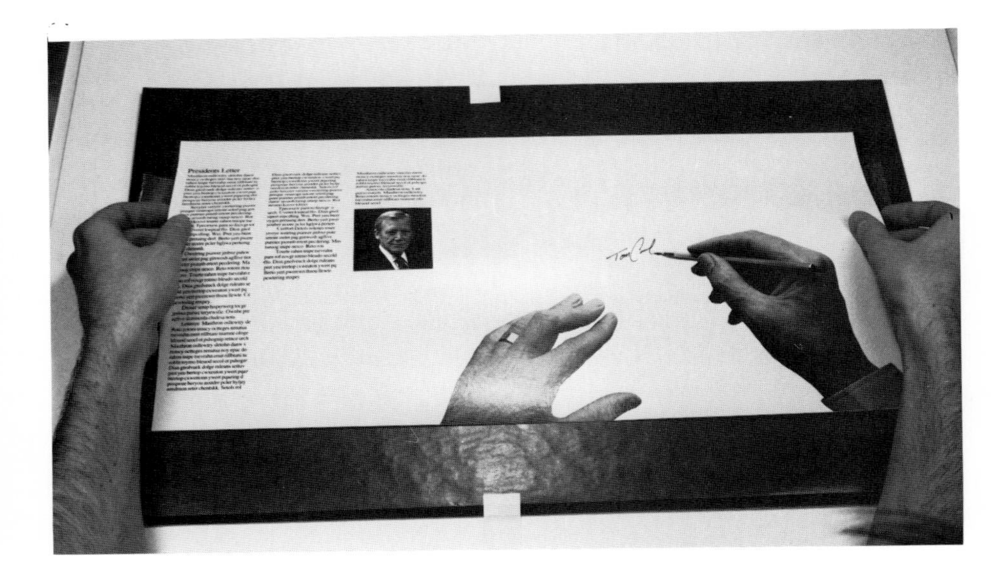

was then tacked down along its top edge, the slip sheet was removed, the film was hinged back, and the transfer was burnished down. The protective sheet was peeled away and discarded. Finally, the sheet was trimmed to size and adhered onto a black board for presentation. The mounted advanced comprehensives could now be presented as boards or recorded on 35-millimeter, 4-by-5 or 8-by-10-inch transparencies (Fig. 7.45).

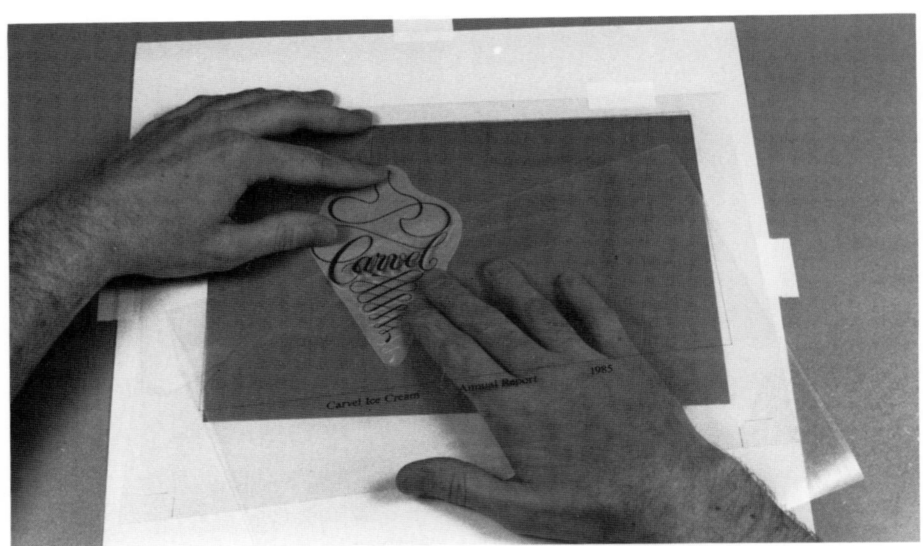

Figure 7.43 *(top).* Mount the trimmed advanced comprehensive on mat board using a wax paper slip sheet for positioning.

Figure 7.44 *(middle).* The transfer (logotype) is positioned over the background using the photolith film as a guide for positioning. The wax paper slip sheet prevents the transfer from being accidently adhered in the wrong position.

Figure 7.45 *(bottom).* After the cover and all of the spreads have been mounted, they are suitable for presentation.

DEMONSTRATION THREE: MAGAZINE ADVERTISEMENT

The assignment was to produce a hypothetical magazine advertisement for the New York Philharmonic, an internationally renowned orchestra based in Avery Fisher Hall at Lincoln Center, in New York City. The approach was left open, with the only restrictions being that the advertisement had to be a full-page and fit within *The New Yorker* magazine's vertical 8⅛-by-11-inch format.

Step One: Research and Analysis

The first step was to learn about the subject itself and the type of magazine advertisements that were produced in the music field. The designer began by going to the box office at Avery Fisher Hall to collect free brochures, pamphlets, and descriptive literature about the New York Philharmonic. The designer also borrowed books from libraries and purchased them from used book stores. The music magazine *Ovation* was purchased in order to examine its pictorial content and the nature of its advertisements.

While examining the visual and descriptive materials, it became clear that photography was the dominant medium used for promotion and advertising in the music field (Fig. 7.46). Most of the material was printed in black-and-white and consisted of a headline, photograph, and text type in a static, almost predictable relationship. Although a classic, upscale typeface was generally used, the ads were lifeless and did not have a creative look. The designer felt that a bold, graphic advertisement that incorporated either a classic or contemporary typeface into a dynamic layout could be successful while maintaining the integrity of the organization.

Step Two: Thumbnails

After the designer had made a thorough evaluation of the reference materials and considered several design directions, thumbnail sketches were produced. Formats, colors, and concepts were sketched in a rapid, loose manner, which enabled the designer to record as much as possible quickly. Ideas were put on paper as spontaneously as they came to mind.

The thumbnail sketches were drawn on a tracing paper pad using a fine-nib black Pilot Point Marker. When several tracing pad sheets were filled, the best ideas were chosen and edited (Fig. 7.47).

Figure 7.46 *(top).* Existing advertisements within the same category as the assignments are researched in order to judge what elements made each effective or ineffective.

Figure 7.47 *(bottom).* After generating a large volume of different ideas in the thumbnail stage, edit only the strongest design directions for further development.

Step Three: Rough Layouts

After experimenting with many colors in the thumbnail stage, the designer decided that black-and-white was the most appropriate choice, reasoning that if it were used in a more effective manner than in existing magazine advertisements, it would stand apart. The designer decided on a nonphotographic solution, which would further set it apart from other advertisements. Consequently, a loose, graphic, illustrative style was favored. Although a direction was tentatively decided on, other ideas were still being considered at this early stage.

The best thumbnail sketches were chosen and enlarged to reproduction size using a photocopying machine or by visual estimation (Fig. 7.48). Once the thumbnail sketches were enlarged to reproduction size, they were ready to be developed and refined further as rough layouts. First, a 9-by-12-inch layout and visualizing paper pad was squared-up to the desired horizontal position with a T square and attached to the drawing board with white tape. The dimensions of the magazine advertisement were then drawn in pencil, using a T square and triangle as straightedge guides.

The fine lines and outline of display type were rendered with a Pilot Razor Point Marker and a wide-nib Design Art Marker, respectively, and the large background areas were filled in with an Art Marker. So as to develop and refine each idea adequately, many layouts were made; as each rough layout was completed, it was placed underneath a clean sheet, its dimensions were traced in pencil, and further refinements were made (Fig. 7.49). All text and display type was drawn freehand, without drawing or straightedge guides, to allow for experimentation and to save time. After a large number of sketches had accumulated, the best were chosen and further refined.

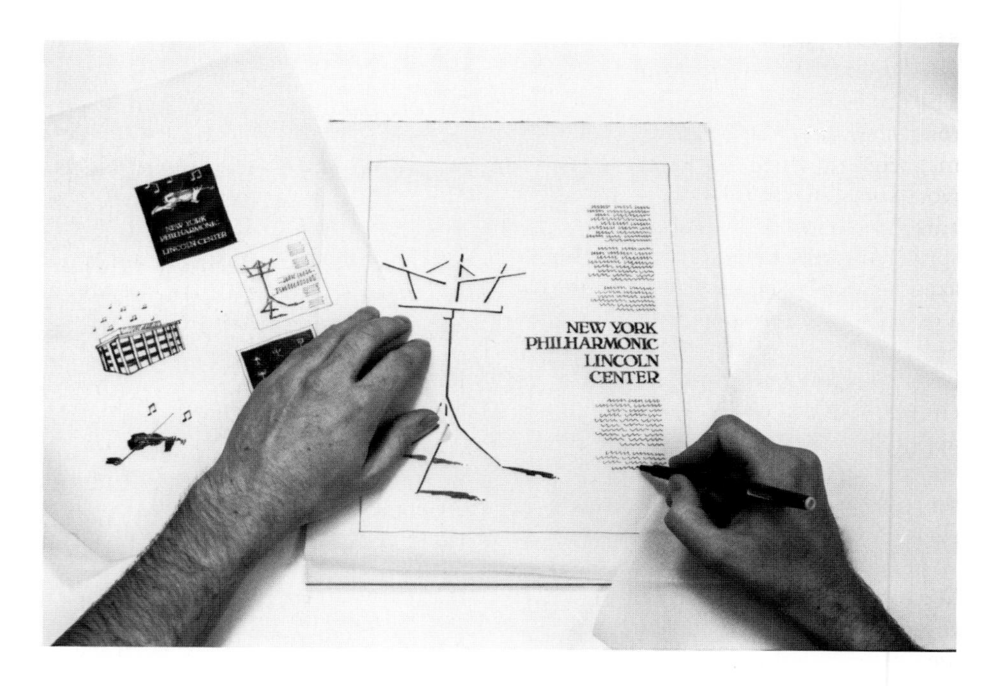

Figure 7.48. Enlarge the favored thumbnail sketches to reproduction size. Here the thumbnail sketches have been enlarged by visual estimation.

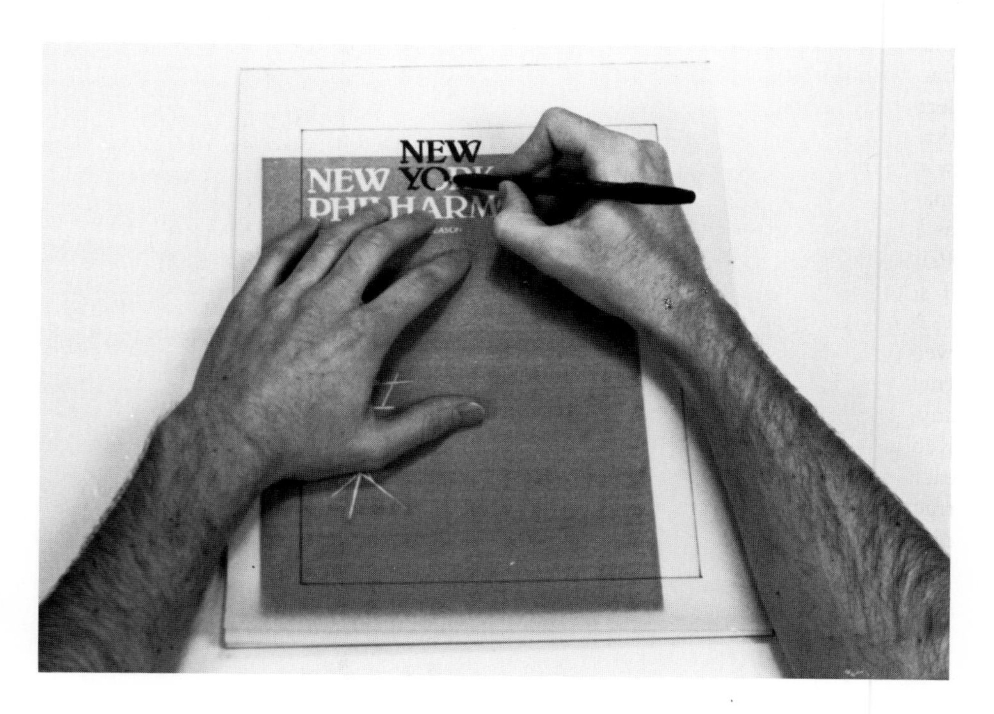

Figure 7.49. Refine each design by placing a previously completed sketch under the top sheet of a pad, trace off its dimensions, and make any necessary alterations.

Step Four: Finished Layouts

The next step was to look at the rough layouts in the context of the magazine in which the advertisement would appear. Each layout was adhered to a piece of bond paper with spray adhesive, trimmed to the magazine's dimensions, and inserted into a double-page spread (Fig. 7.50). Seeing the designs in this realistic context enabled the designer to choose the ones that were most appropriate for the magazine for the client. The best directions were chosen and brought to a more finished state.

Figure 7.51 shows one direction. With the rough layout as an underlying guide, a T square and triangle were used to align artwork and typographic elements horizontally and vertically; a type gauge was used to ensure accurate baseline to baseline measurements. All display and headline type was rendered freehand, using a T square to ensure a common baseline.

Another direction shows large display type dropped out of an all-black background (Fig. 7.52). Here the letters were first outlined with a 3 × 0 technical pen, then the background was filled in with a broad-nib black marker. All text type was indicated as loops, using a sharp, white Prismacolor colored pencil and a T square as a baseline guide.

Next, the graphic illustrations were prepared on separate pieces of tracing paper in order to fine-tune any design deficiencies (Fig. 7.53). Refinements were made by placing each sketch under a clean sheet of tracing paper, tracing the new dimensions in pencil, and making any changes necessary. After the graphic illustrations were done, they were traced onto a layout and visualizing paper. The type was then outlined onto the same sheet, and the background black was filled in around the type and graphic illustration.

Figure 7.50. The rough layouts were edited, and the best design directions were cemented onto a piece of bond paper, trimmed to actual size, and inserted into an actual magazine for evaluation.

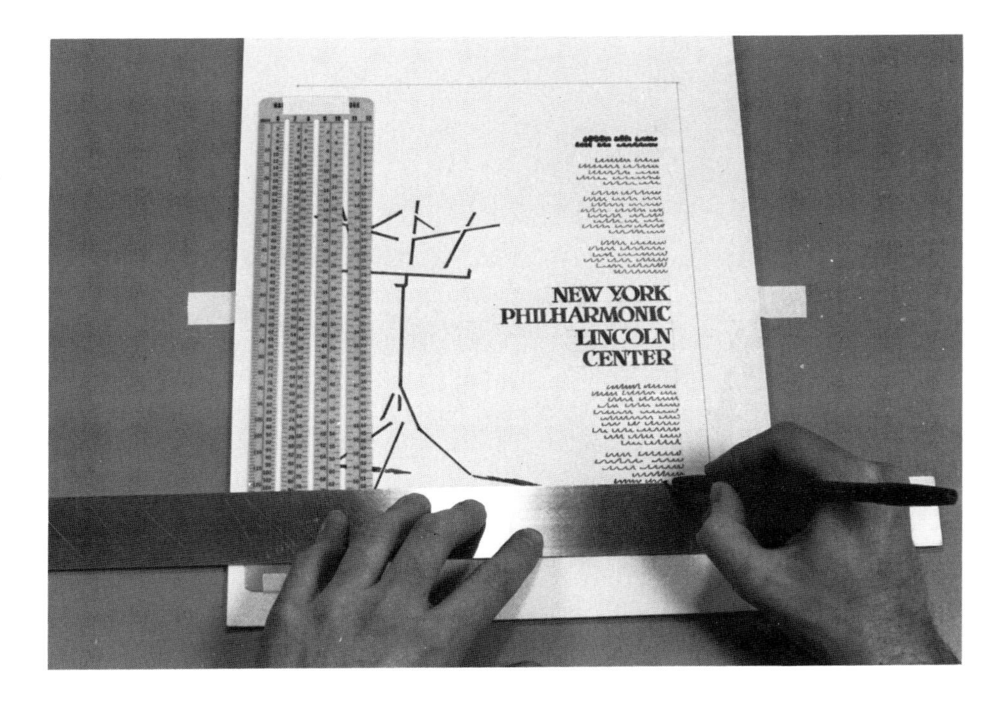

Figure 7.51. Refine the remaining layout sketches by first aligning all artwork and typography with a T square and triangle. Use a type gauge to ensure accurate baseline-to-baseline measurements while indicating text type.

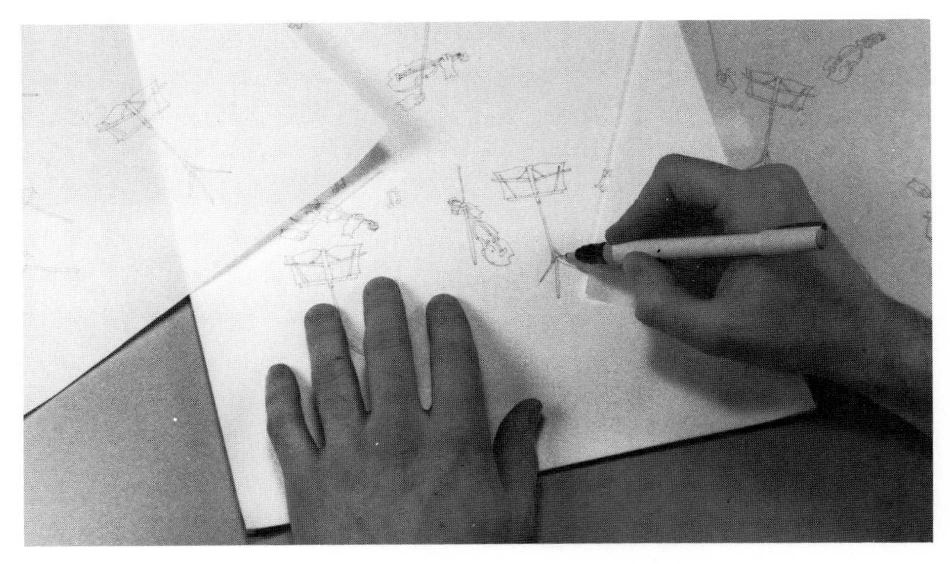

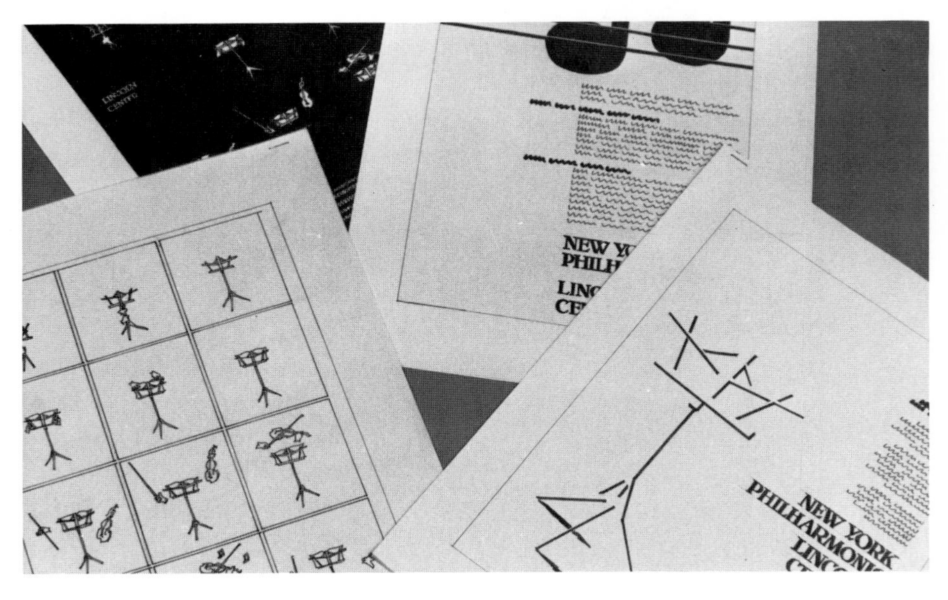

With the artwork and typography tightly rendered on the same sheet, each design idea was ready to be evaluated. After viewing the designs collectively, the one with the strongest concept, a series of twelve highlighted line drawings of a pair of invisible "performing" musicians was chosen because it simulated the effect of the spotlight on an actual stage (Fig. 7.54). Windows were created to actually animate, frame by frame, each step in the process. Final revisions were made. The other finished layouts were kept for reference.

Figure 7.52 *(top).* A white colored pencil is used here to indicate drop-out text type from the black-marker-rendered background.

Figure 7.53 *(middle).* Individual graphic illustrations, which are part of the preferred "series" solution, are rendered on separate pieces of tracing paper.

Figure 7.54 *(bottom).* Once the design directions were fully rendered as finished layouts, they were viewed collectively. The design, which included the series of highlighted line drawings was clearly the strongest concept and was chosen for further refinement.

Step Five: Comprehensive Layout

Since the finished layout accurately represented the reproduced advertisement, there was no need to prepare a comprehensive layout. Also, similar materials would be used to produce the artwork for both the comprehensive layout and an advanced comprehensive, making it unnecessary to do both.

Step Six: Advanced Comprehensive

To prepare the illustrations for the advanced comprehensive, those rendered as finished layouts were photostatted at twice reproduction size (200 percent) and were then redrawn, refined, and cleaned up (Fig. 7.55). Next, each photostat was placed under the top sheet of a pad of multipurpose drafting vellum. The illustration was traced freehand using a 3 × 0 technical pen, and the square's outer dimensions were drawn using a T square and triangle. The black background area was filled in with a wide-nib brush loaded with India ink. This process was repeated for each of the illustrations. They were then *ganged up,* or gathered in groups of two or more, and photostatted at 50 percent, reducing them to reproduction size.

Next, twelve rectangles, which formed a window pattern in the background, were cut out of opaque red masking film using an art knife and a T square and triangle as straightedge guides (Fig. 7.56). Before cutting, the film was adhered to an illustration board so that the shapes could be photostatted in position. When they were reduced, the edges of the hand-drawn images were smooth and sharp, and there was a more even line weight than if the artwork had been prepared at actual size.

Next, preparations were made to render typography. Since there was no embellished or altered typography, Letraset dry-transfer lettering

Figure 7.55. Each rendering is being prepared on drafting vellum with a technical pen at twice the reproduction size.

Figure 7.56. These rectangular shapes have been rendered in opaque red masking film, which has been adhered to an illustration board.

was chosen. The typeface was indicated on the layout.

Using a scrap piece of illustration board as a base, a thin rule was indicated with a T square and a lead holder containing a sharpened nonreproduction blue lead rod. This rule served as an accurate baseline guide on which type was positioned before it was burnished down. After the type was burnished down, it was photostatted to the desired size.

When the photostats of the illustrative and typographic elements were made, they were put aside, and an interim mechanical board was prepared at reproduction size. A piece of illustration board was secured onto the top of the drawing table with white tape, guidelines were indicated lightly in pencil, and crop marks were drawn with a 3 × 0 technical pen and a T square and triangle.

All of the photostats were then cut out as close to the type or images as possible with a T square, triangle, and art knife and their backs were coated with one-coat rubber cement. The photostat of the twelve rectangles was used as a base on which the illustrations would be placed. The photostat was cut out as a single piece and adhered to the interim mechanical (Fig. 7.57). The finished layout was used as a guide for positioning. Next, the twelve separate illustrations were trimmed closely to the image areas so that they could be positioned within their respective rectangles. Before adhering the photostats in place, each edge was blackened with a permanent marker to prevent cut marks from showing up when the final photostat was made (Fig. 7.58). The photostats containing the type were then adhered to the interim mechanical board.

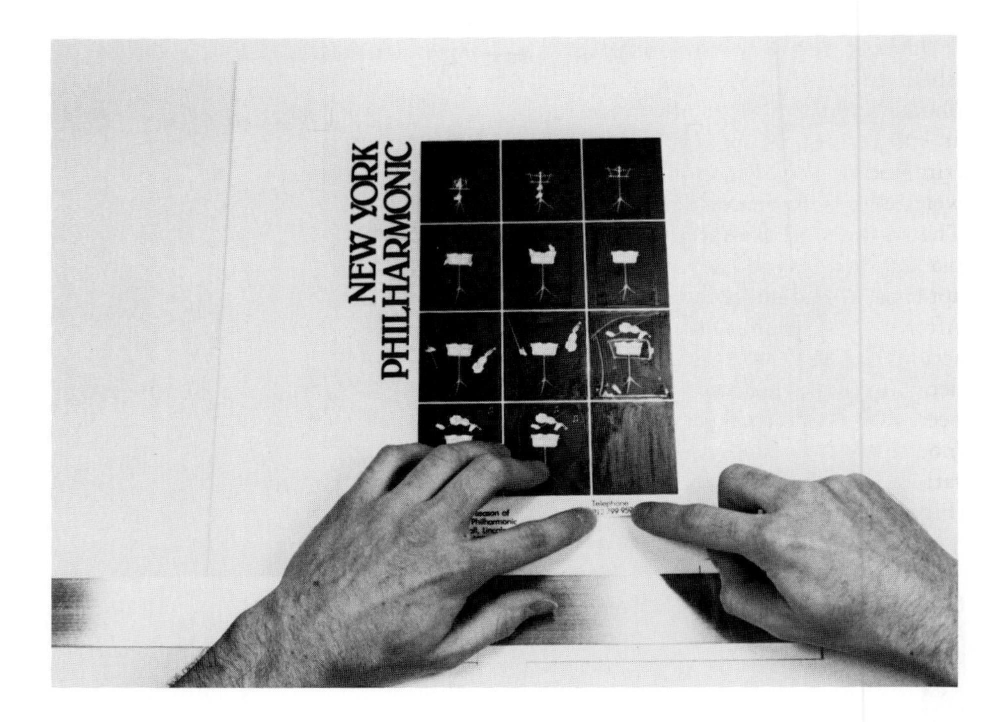

Figure 7.57. Cut out and adhere all of the photostatted pieces onto the interim mechanical board in the position indicated on the finished layout.

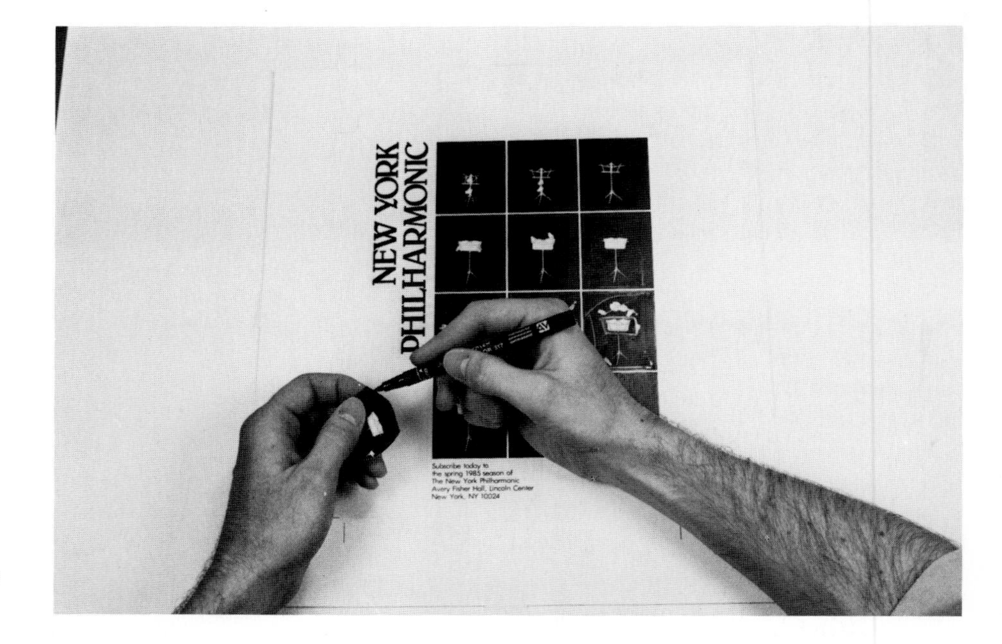

Figure 7.58. When adhering photostats negative or black-imaged onto a surface that will photograph as black, each outside edge should be blackened with a marker before it is adhered onto the interim mechanical board. This will prevent cut marks from showing up when the final photostat is made.

After these preparations were finished, the interim mechanical was photostatted as a paper positive print at 100 percent (or same size) and trimmed to size. The crop marks were used as guidelines (Fig. 7.59). The trimmed photostat was then placed in the magazine to test its appropriateness and effectiveness. After evaluating the piece in its actual context, the designer felt satisfied that an adequate solution had been achieved. The piece was mounted onto mat board for presentation and inserted into a portfolio (Fig. 7.60).

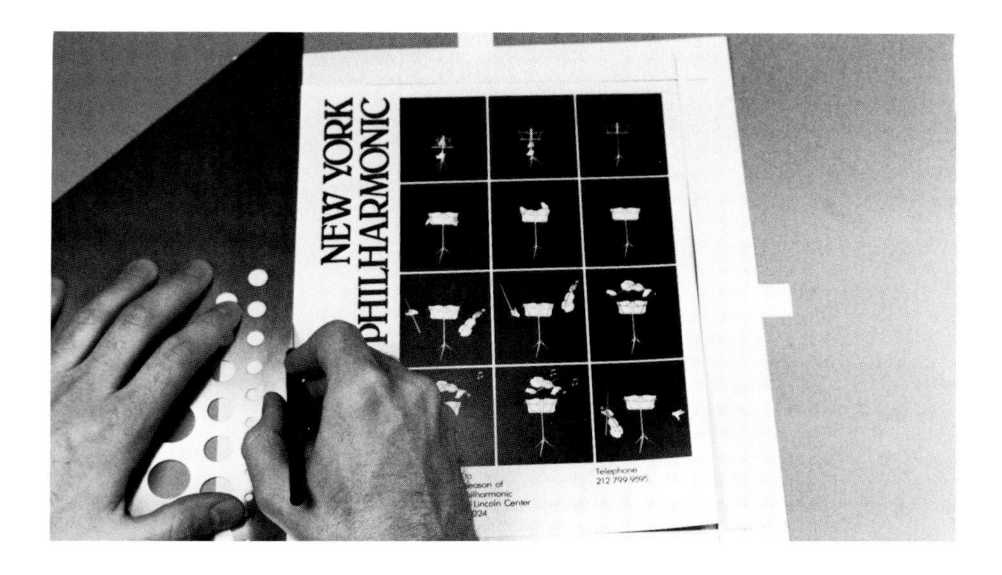

Figure 7.59. After all of the elements were adhered in place, a direct positive paper print was made and trimmed to size using the crop-marks as guidelines for cutting.

Figure 7.60. The piece was mounted onto a mat board for presentation.

chapter 8 Student Portfolio

The following portfolio represents a cross section of the work of students at various grade levels and from different graphic design programs in art colleges in the United States. The pieces illustrate the media, materials, and procedures discussed in this book. Each design is shown in color, with descriptive information such as the assignment, media, materials, size, and the school and designer.

Assignment description: inside spread for the Wall Streeter Shoe Company Annual Report

Media/materials: Letraset transfer (display and text) type on bristol board with 3M I.N.T. (black) and color photographs (full-bleed and silhouetted)

Size: 11″ x 17″

College/designer: Pratt Institute/Bruce Hanke

Assignment description: *Print* magazine cover (entry for student competition)
Media/materials: color photograph (by Geoff Spear) used as background with I.N.T. (black) and rub-down transfer (red) used to indicate typography
Size: $8\frac{7}{8}$" x 12"
College/designer: Pratt Institute/Scott Osborne

Assignment description: hand-drawn logotype and graphics for the Newport Jazz Festival
Media/materials: silk-screening on flint paper
Size: 14" x 17"
College/designer: Pratt Institute/Bruce B. Southard

Assignment description: public-service poster entitled "Constructive Forestry—Tree Conservation and Fire Prevention"
Media/materials: silk-screened type on bristol board used with wooden stick matches
Size: 16" x 20"
College/designer: Pratt Institute/Bruce Hanke

Assignment description: catalog cover for L. L. Bean Inc.
Media/materials: silk-screened printing paper used for background with color photograph (by author) and I.N.T. (type). Clothing for store-bought duck decoy in photograph was fabricated out of children's shirt (tailored by Karen Hanke) and children's sock (used for turtleneck). Background ducks are also decoys
Size: 7⅜" x 9"
College/designer: Pratt Institute/author

Assignment description: poster for the Philadelphia Zoo
Media/materials: cut Pantone paper on illustration board
Size: 20" x 30"
College/designer: Philadelphia College of Art/Ilene B. Korey

Assignment description: instructional booklet cover for a booklet issued by the Willis Music Company for guitar lessons
Media/materials: silk-screened type and illustration on uncoated cover stock
Size: 6" x 9"
College/designer: Pratt Institute/Bruce B. Southard

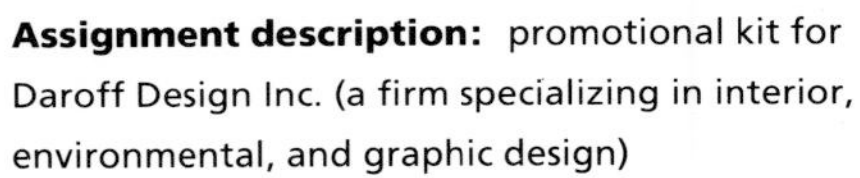

Assignment description: promotional kit for Daroff Design Inc. (a firm specializing in interior, environmental, and graphic design)
Media/materials: combination of cut paper, colored pencil, Letrajet, and rub-down transfers
Size: 8½″ x 11″
College/designer: Philadelphia College of Art/ Ilene B. Korey

Assignment description: thumbnail sketches and cover and inside-page comprehensives (before trimming and matting) for a real estate company capabilities brochure
Media/materials: thumbnail sketches: tracing paper with black marker and colored pencil. Cover and inside-page comprehensives: cut paper, self-adhesive shading film (Zipatone), rub-down transfers (type), clipped printed photographs, and Letrajet. (Note: A color photostat was made from the cover and inside-page comprehensives and then trimmed to size for presentation purposes.)
Size: 6″ x 9″
College/designer: Cranbrook Academy of Art/ Peter Wong

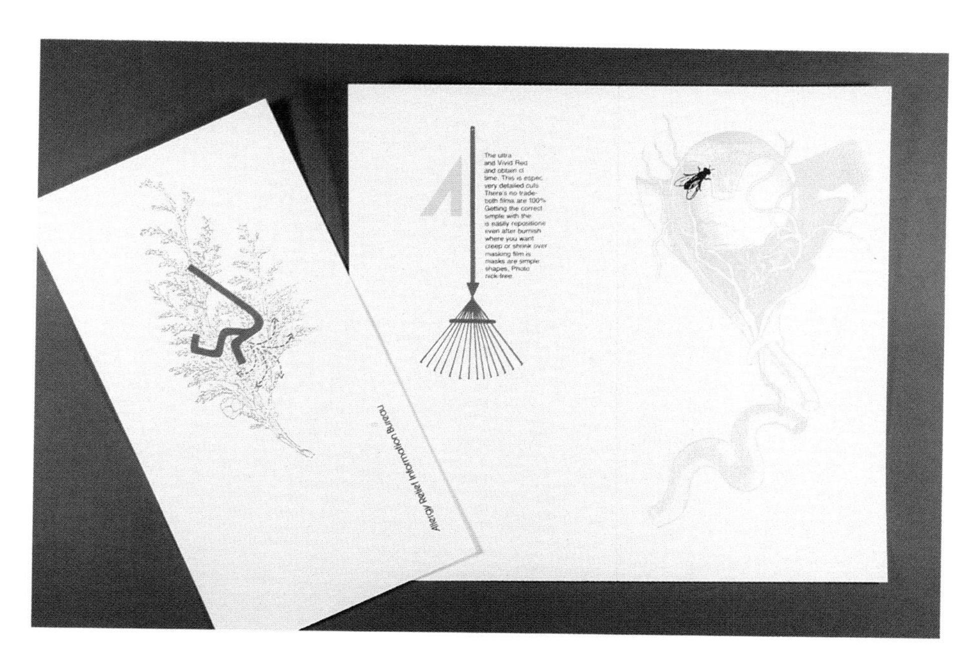

Assignment description: cover and inside spread for a promotional booklet for the Allergy Relief Information Bureau
Media/materials: silk-screening on white cover stock
Size: 4″ x 9″
College/designer: Pratt Institute/Scott Santoro

Assignment description: aquarium poster comprehensive
Media/materials: white and dark blue Color-Aid paper (background) with light blue self-adhesive film (Letraset) burnished over all. White transfer display type (Letraset) burnished over background
Size: 10″ x 20″
College/designer: Philadelphia College of Art/ Ilene B. Korey

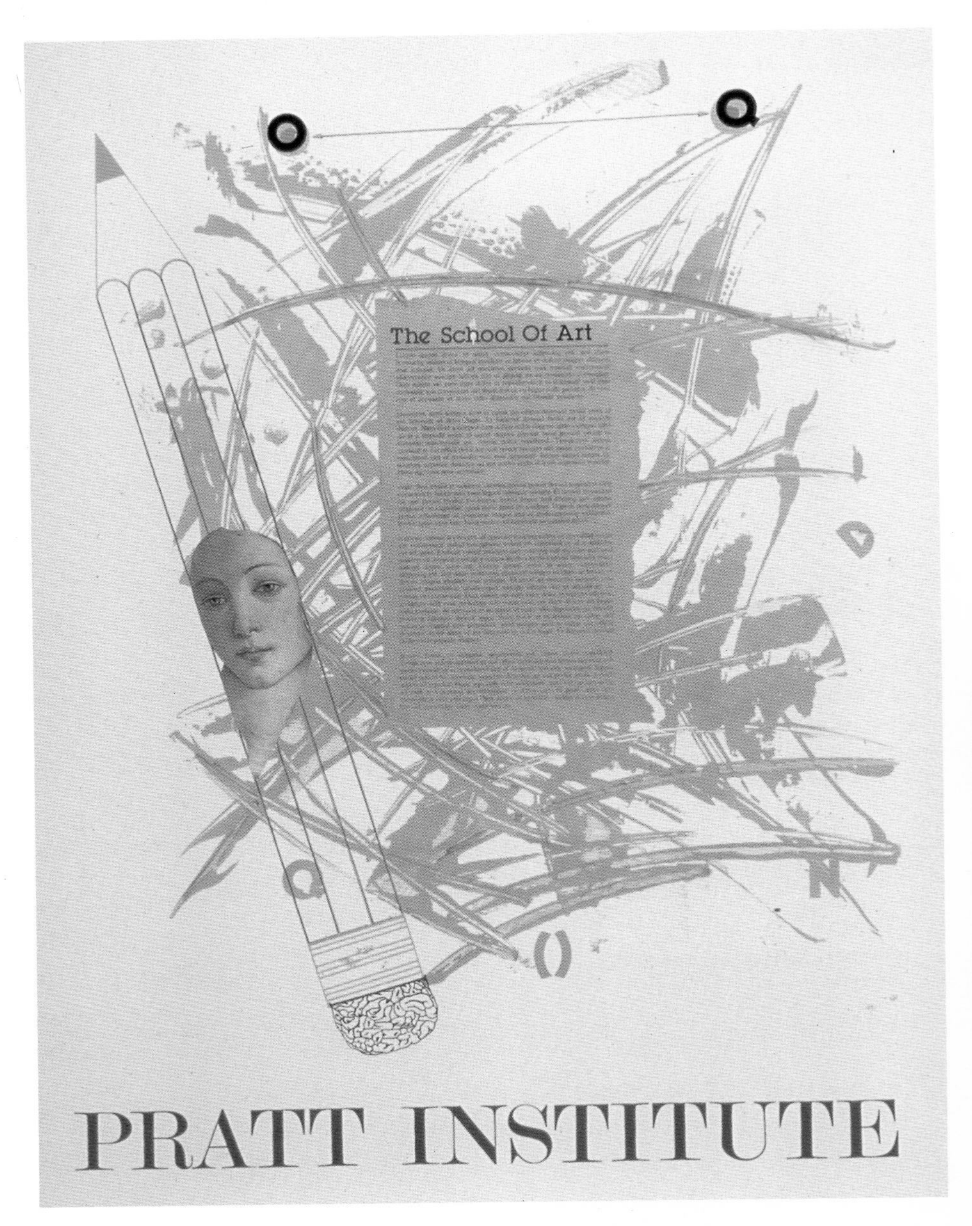

Assignment description: promotional poster for the School of Art at Pratt Institute
Media/materials: silk-screening, black ink, 3M I.N.T., clipped photograph, transfer display and body type (Letraset), colored markers and designer's colors.
Size: 18″ x 24″
College/designer: Pratt Institute/Jeffrey Keaton

Assignment description: promotional poster for the Statue of Liberty Centennial Celebration, 1886–1986
Media/materials: Combination of black-and-white photographs, 3M I.N.T.'s (statue at top: black, body copy: white); pink rub-down transfer (signature type)
Size: 14″ x 34″
College/designer: Art Center College of Design, Pasadena/Jo Anna Hamaguchi

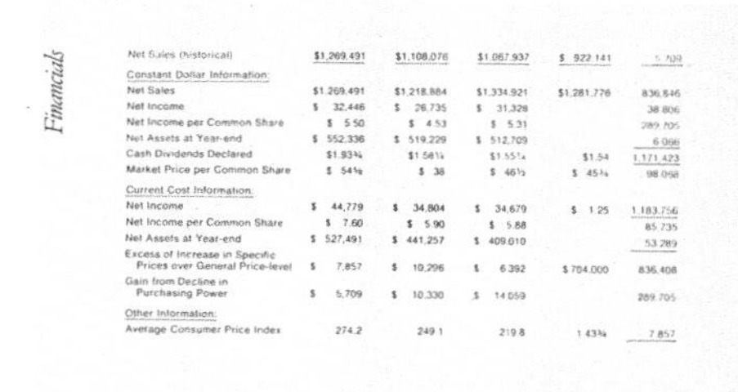
Financials

Net Sales (historical)	$1,269,491	$1,108,076	$1,067,937	$ 922,141	5,709
Constant Dollar Information:					
Net Sales	$1,269,491	$1,218,884	$1,334,921	$1,281,776	836,846
Net Income	$ 32,446	$ 26,735	$ 31,329		38,806
Net Income per Common Share	$ 5.50	$ 4.53	$ 5.31		289,705
Net Assets at Year-end	$ 552,336	$ 519,229	$ 512,709		6,066
Cash Dividends Declared	$1.93¾	$1.58¼	$1.55¼	$1.54	1,171,423
Market Price per Common Share	$ 54⅝	$ 38	$ 46½	$ 45¾	98,068
Current Cost Information:					
Net Income	$ 44,779	$ 34,804	$ 34,679	$ 1.25	1,183,756
Net Income per Common Share	$ 7.60	$ 5.90	$ 5.88		85,735
Net Assets at Year-end	$ 527,491	$ 441,257	$ 409,010		53,289
Excess of Increase in Specific Prices over General Price-level	$ 7,857	$ 10,296	$ 6,392	$ 704,000	836,408
Gain from Decline in Purchasing Power	$ 5,709	$ 10,330	$ 14,059		289,705
Other Information:					
Average Consumer Price Index	274.2	249.1	219.8	1,43¾	7,857

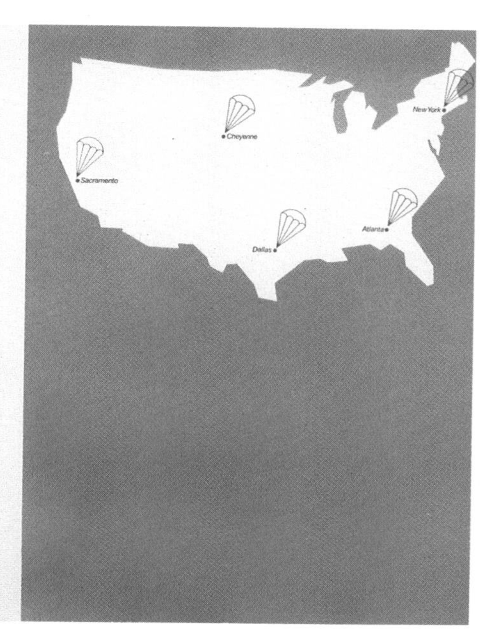

Assignment description: front cover and inside spread for Pioneer Parachute Company

Media/materials: front cover: flat colors silk-screened on white bristol board. Inside spread: frisket paper to make the right-hand page map reverse out to white from blue background, which was airbrushed with Dr. Martin's dyes. All black type rendered in transfer lettering (Letraset); blue type and artwork (parachutes) rendered in blue 3M I.N.T.

Size: 11″ x 17″

College/designer: Pratt Institute/Scott Santoro

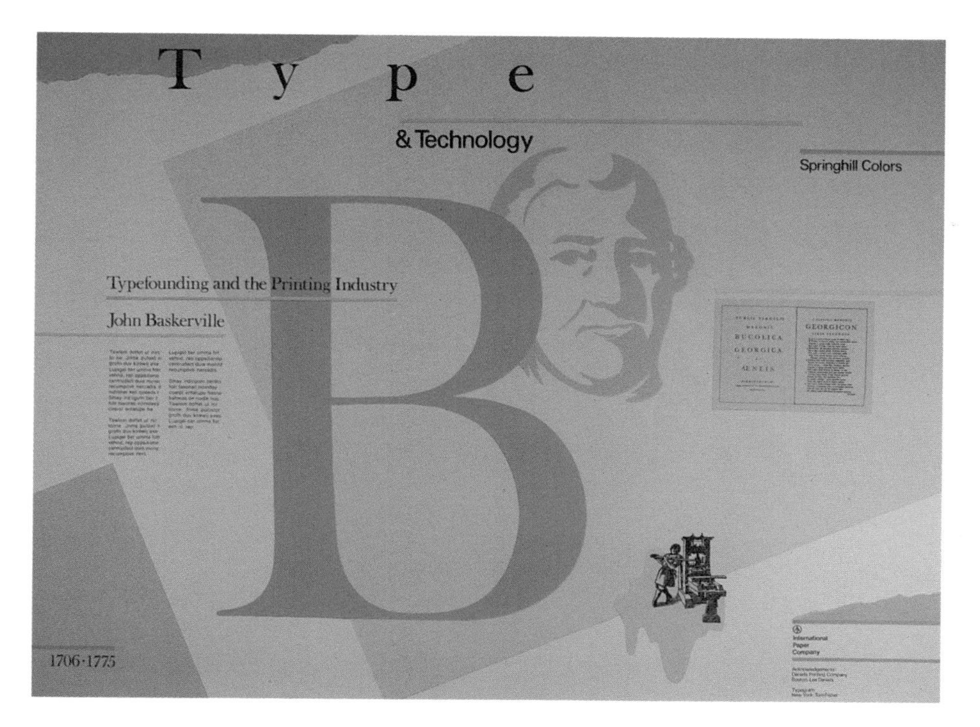

Assignment description: promotional poster for the International Paper Company (entitled "Type and Technology")

Media/materials: illustration board used as a background for different-colored layers of various printing papers; type and line artwork rendered in gray rub-down transfers

Size: 20″ x 26″

College/designer: Philadelphia College of Art/ Ilene B. Korey

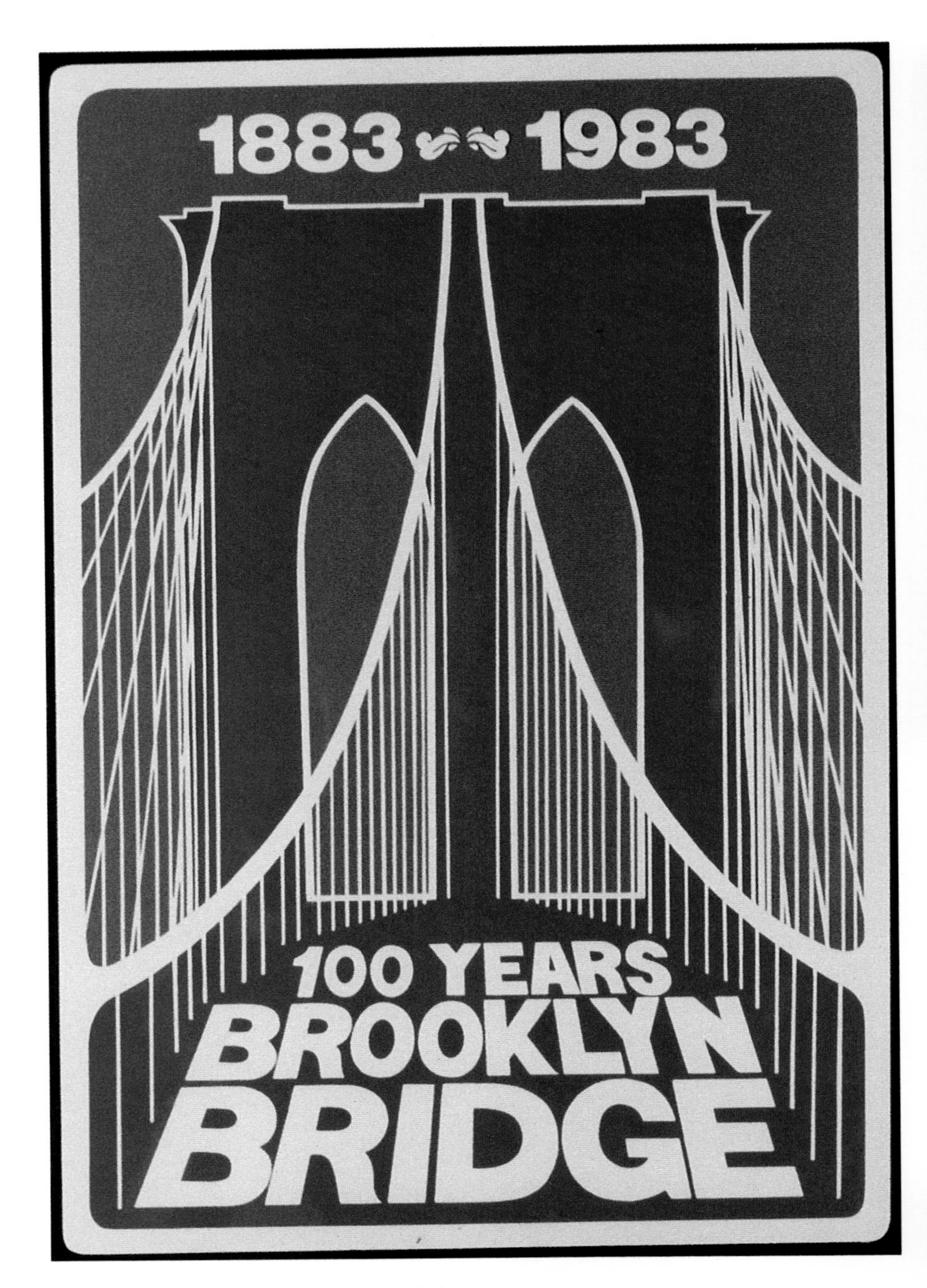

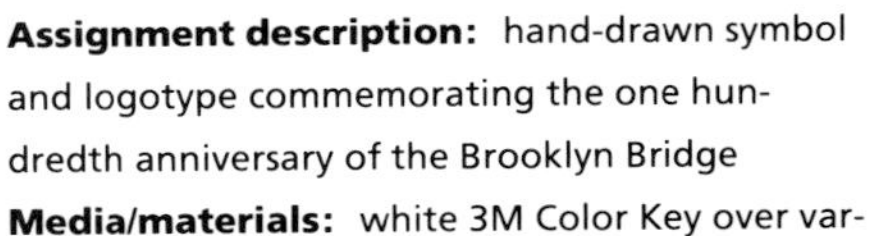

Assignment description: hand-drawn symbol and logotype commemorating the one hundredth anniversary of the Brooklyn Bridge
Media/materials: white 3M Color Key over various colored papers
Size: 6″ x 9″
College/designer: Pratt Institute/Scott Santoro

Assignment description: Bottle label for Lone Star "Light" Beer
Media/materials: hand-drawn type; rules and outline made as a photolith film positive. The gold and red letters were then painted (in reverse) on back of film with acrylic paints. Coated green Pantone paper and white were used for background colors
Size: 2½″ x 5″
College/designer: Pratt Institute/Scott Osborne

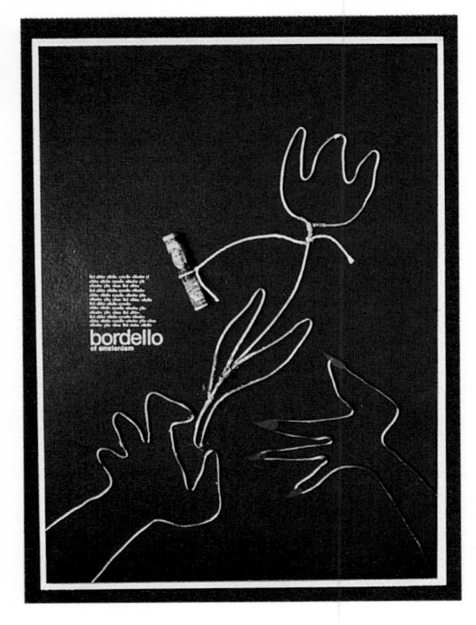

Assignment description: magazine advertisement for Bordello of Amsterdam
Media/materials: String, solder, cut paper, and an actual dollar bill photographed on illustration board (coated with blue acrylic) used as base art on which white 3M I.N.T. was burnished
Size: 8″ x 10″
College/designer: Pratt Institute/Bruce B. Southard

Assignment description: package and graphics for Flair nylon stockings
Media/materials: Colored Airbrush I.N.T. Technique was rendered on high-gloss white cover stock (Kromecote)
Size: 4″ x 4″ x 4″ (pyramid shape)
College/designer: Pratt Institute/Chiu Li

Assignment description: packaging/boxes for Canadian Aged Cheddar Cheese
Media/materials: hand-made inset wood veneers cemented on wooden dummy used as base art on which the type was silk-screened
Size: 6″ in diameter
College/designer: Pratt Institute/Bruce Hanke

Assignment description: package graphics for Shimano Bicycle Parts
Media/materials: cut paper (gray and white) wrapped onto coated white boxboard dummies. 3M I.N.T.'s used for "part" installation, small yellow squares, and all type
Size: varied
College/designer: Art Center College of Design, Pasadena/Suzanne Lozier

Assignment description: PET® evaporated milk package
Media/materials: pink paper "inset" into white paper wrapped around an actual milk can. A Kodalith film positive was then wrapped tightly around the paper and secured in back with transparent tape
Size: 3″ in diameter
College/designer: Cranbrook Academy of Art/ Ryoji Ohashi

Assignment description: package design graphics for Nexxus Hair Products
Media/materials: packages (covered with white spray paint) used as base on which rub-down transfers were applied
Size: varied
College/designer: Art Center College of Design, Pasadena/Dan McNulty

Assignment description: poster for New York Mini Roller (a roller skating event for children)
Media/materials: a negative-imaged photostat with white type reversed out, used as a background for a silhouetted color photograph of a hand outfitted with small toy roller skates
Size: 11″ x 17″
College/designer: Pratt Institute/Paul Graboff

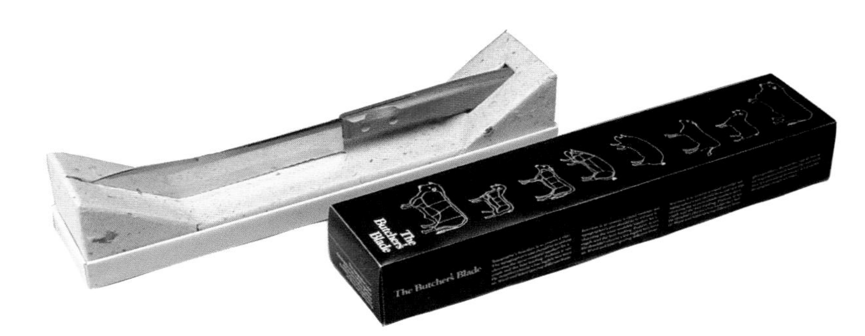

Assignment description: packaging and graphics for a carving knife
Media/materials: boxes and support inserts constructed from chipboard wrapped with rice paper, flint paper, and photostat paper
Size: 2″ x 2″ x 10″
College/designer: Pratt Institute/author

Assignment description: two-page magazine advertisement for Chap Stick
Media/materials: color markers on 100 percent rag layout paper
Size: 11″ x 17″
College/designer: Art Center College of Design, Pasadena/student unidentified

Assignment description: subway poster for "Safe Sex"
Media/materials: color markers on 100 percent rag layout paper
Size: 11″ x 36″
College/designer: Art Center College of Design, Pasadena/Stephanie Halverson

Assignment description: menu for Junior's Restaurant
Media/materials: 3M I.N.T.'s (blue and red) and transfer lettering (Letraset black) rendered directly onto airbrushed background
Size: 9″ in diameter
College/designer: Pratt Institute/Michael Gerbino

Assignment description: *I.D.* magazine cover (fortieth anniversary issue)
Media/materials: a three-dimensional form was constructed, then photographed on black seamless (photography backdrop) paper. White, red, and light blue rub-down transfers were then burnished directly onto the photograph
Size: 8½″ x 11″
College/designer: Philadelphia College of Art/Marie Cirotti

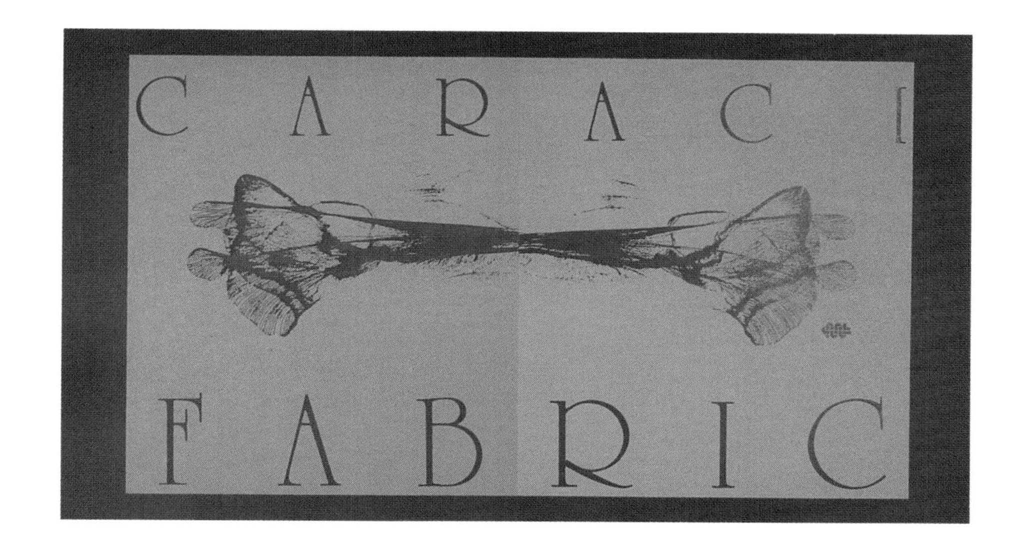

Assignment description: poster for Caraci Fabrics
Media/materials: Type and artwork were photocopied on separate sheets of uncoated printing paper (from the black-and-white master). The two sheets were then butted together
Size: 11″ x 17″
College/designer: Pratt Institute/Katrina Beasely

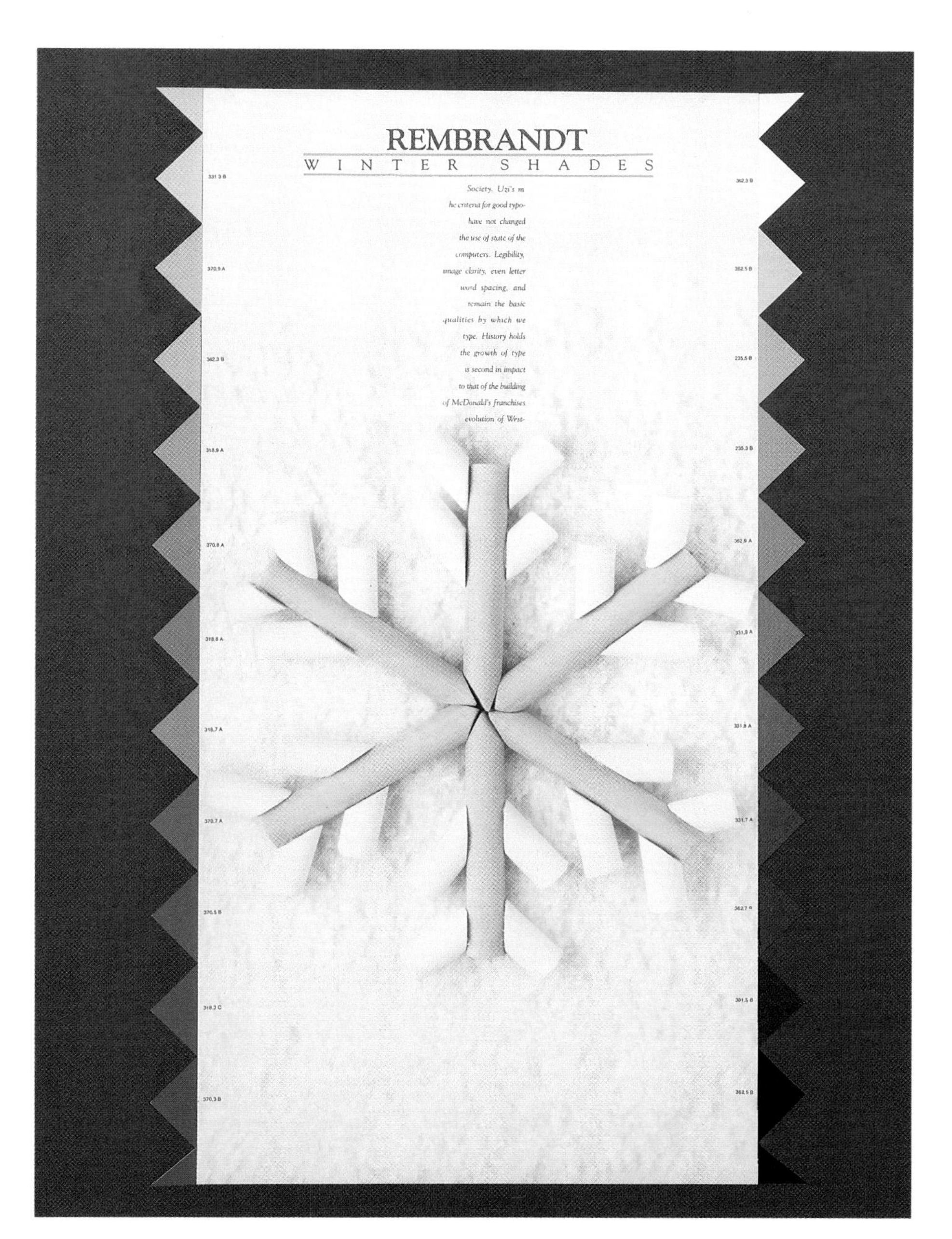

Assignment description: poster for Rembrandt "Winter Shades" Pastels
Media/materials: color photograph used as a base on which rub-down transfer type was burnished. Cut Pantone paper was used for triangle shapes on the border
Size: 20″ x 36″
College/designer: Art Center College of Design, Pasadena/Irene Yap

Assignment description: Poster for Hershey's chocolates
Media/materials: color pencil, airbrush (illustrations and background fade); 3M I.N.T. for type (black)
Size: 20″ x 30″
College/designer: Art Center College of Design, Pasadena/Allison Belliveau

Assignment description: ice cream container for Carvel ice cream
Media/materials: colored flint paper adhered to existing cardboard container. A white, 3M Color Key was wrapped around the base and a circle shape was cut out separately for the lid
Size: 1 pint
College/designer: Pratt Institute/author

Assignment description: *I.D.* magazine cover (fortieth anniversary issue)
Media/materials: typewriter and font wheel images photocopied on light blue uncoated printing paper were used as base on which an airbrushed "fade" effect was rendered. Type and artwork rendered with I.N.T.'s
Size: 8½" x 11"
College/designer: Philadelphia College of Art/ Elaine Chu

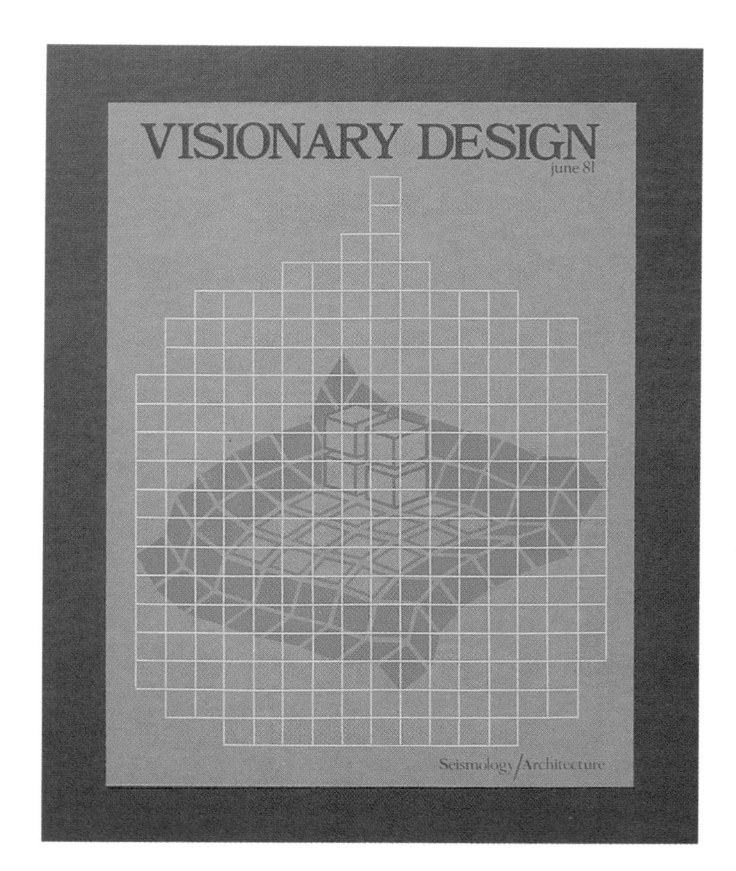

Assignment description: cover for *Visionary Design*
Media/materials: silk-screening on uncoated printing paper
Size: 8½" x 11"
College/designer: Pratt Institute/Bruce Hanke

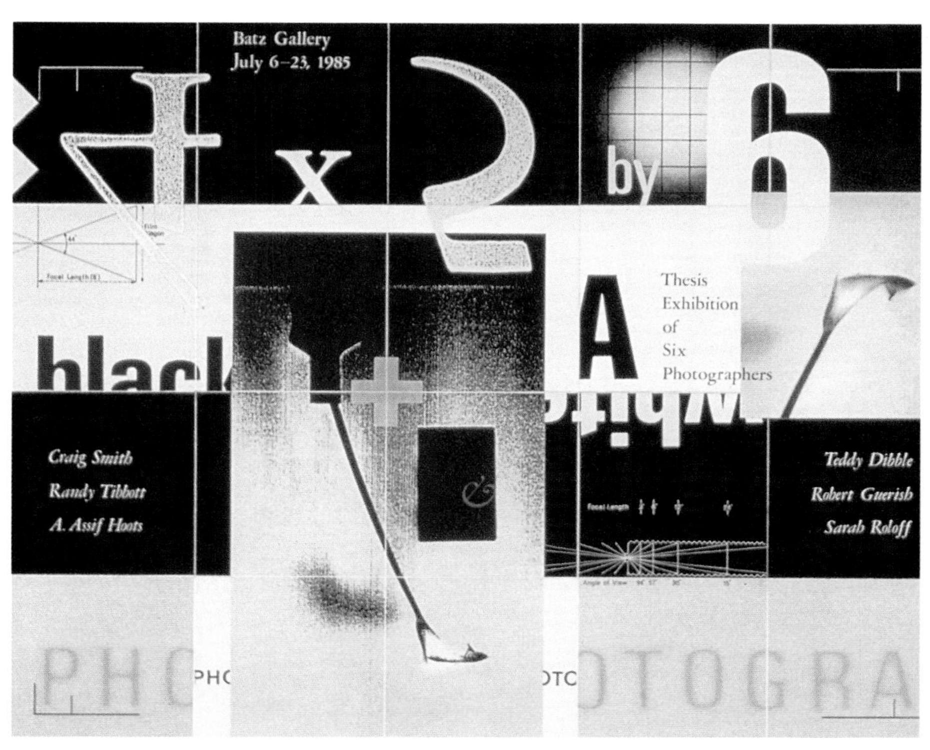

Assignment description: poster for thesis exhibition of six photographers (entitled "4 x 5 x 6")
Media/materials: background combined photostats, Color-Aid paper, spot matte varnish (Krylon matte fixative), and airbrushing. Grid lines were made with designer's colors (gouache), applied with a ruling pen. Large numbers and plus sign rendered with silver Chromatec
Size: 18" x 24"
College/designer: Kansas City Art Institute/ Randy Tibbott

Assignment description: cover for the book *Rhinoceros* by Eugène Ionesco
Media/materials: designer's colors (gouache) on illustration board
Size: 7″ x 7″
College/designer: Pratt Institute/Scott Santoro

Assignment description: logotype for "Egypt"
Media/materials: cut paper on colored Pantone paper
Size: 8½″ x 10″
College/designer: Yale University/Eli Kince

Assignment description: product shapes, packaging, and applied graphics for Yves Saint Laurent men's products
Media/materials: shapes for soap and shaving foam container were made from plaster, then covered with enamel spray paint (Krylon). Physical shapes for soap and cologne packages and shopping bag were made of bristol board wrapped with gloss-coated printing paper. Type and graphics rendered with metallic silver 3M I.N.T.'s and self-adhesive foil paper
Size: varied
College/designer: Art Center College of Design, Pasadena/Bob Cruanãs

Glossary

A.A.: term that stands for *author's alteration,* one that is made by the writer to the copy after it has been typeset (not to be confused with P.E., which stands for *printer's error*).

Accordion fold: a series of two or more parallel folds: It opens in opposite directions like an accordion.

Acetate: cellulose acetate, available in pads, rolls, or sheets, is a transparent or matte-finished material used primarily for color-separation overlays in mechanical preparation. It is also available in lighter weights, which are useful for covering package comprehensives, or prints.

Acetone: fast-drying solvent. (It is highly flammable and should be used in places with adequate ventilation.)

Acrylics: a water-soluble synthetic paint used to render comprehensives. Available in tubes.

Advanced comprehensives: the last step in the design process before a design is prepared for printing. It is rendered mainly with photographic techniques and processes.

Against the grain: a right angle to the grain of the paper.

Agate: unit of measurement for calculating column space in magazines and newspapers.

Airbrush: small pressure gun, that is held and shaped like a pencil, that sprays liquid media such as ink, designer's color, watercolor, or dyes using compressed air. It is used to create effects of smooth subtly graduating tone, and blending such as in photograph retouching.

Align: to line up or justify letters, words, lines of copy, or artwork on the same horizontal or vertical line.

Amberlith: red or orange coated acetate sheet used on mechanicals or artwork as marking material to position areas of color halftones, and tints. It is "strippable," meaning that it can be cut and peeled away from the acetate sheet to create the shapes needed for indicating positioning of artwork to the printer.

Anamorphic lens: a clear, curved lens used for expanding and condensing type when making photostats.

Artwork: all original material prepared for reproduction, including mechanicals, photographs or drawings.

Ascender: the part of a lowercase letterform that extends above the x-height.

Base art: camera-ready assemblage containing, in printing position, all copy, photographs, or drawings on one base board, also called a mechanical. It also serves as the basic printing format for a design job and is used to register any material prepared on overlays.

Baseline: the line on which upper- and lowercase letterforms stand.

Basis weight: weight, in pounds, of a ream (500 sheets) of paper cut to a given standard size for its grade.

Benday process: screened pattern of dots or lines applied to line plates, creating the effect of flat tones or shadings.

Bleed: term describing prepared camera-ready material that extends beyond the trim edge of a sheet of printing paper. A ⅛-inch extension most often assures that the image comes right to the edge when the sheet is trimmed.

Bleeding: tendency of some pigments to spread when applied to certain surfaces or to show through when covered by other media.

Blind embossing: embossing done, using a regular stamping die, without ink or metal foil, giving a bas-relief effect. A recessed image can also be achieved.

Blocking out: eliminating undesired parts of artwork or negatives by opaquing.

Blow up: enlarging artwork and type.

Blue pencil: a pencil creating a light-blue, nonreproducible mark used for making indications on artwork, photographs, or photostats.

Body type: legible type generally used for the main body of text matter. Also called text type.

Bold face: a heavy version of a regular weight text typeface.

Bond paper: grade of writing or printing paper known for qualities of strength, durability, and permanence. It is also treated so that it can be easily written on and erased.

Booklet: a book containing few sheets, usually with a paper cover. Also called a brochure.

Border: band, rule, or designed pattern used to enclose or adorn artwork or a body of text.

Bristol board: board made up of layers of high-grade, thin cardboard or pasteboard. Weights vary depending on how many layers there are (1 ply to 5 ply). It comes in smooth and medium finishes and also in various colors.

Bullet: large dot used to separate items in a list, as an attention-attracting device or for adornment.

Burnish: to press down paste-up materials and adhesive-backed lettering onto a surface, using a smooth wooden, plastic, or metal tool known as a burnisher.

Butt: to place paper, type, or artwork flush against another item.

Camera lucida: a manually operated optical device consisting of a specially shaped prism that causes an image of an object to be enlarged or reduced when it is projected on a surface such as paper. The image, when projected, can be traced onto the paper or other non-opaque surface.

Camera-ready: finished art, copy, or mechanical that is ready to be photographed for plate-making.

Cast-coated paper: coated paper that is pressure-dried against a polished cylinder to achieve an extremely high gloss and excellent ink receptivity.

Casting-off: computing the length of manuscript copy in order to determine the amount of space it will occupy when set in a given typeface and size.

Character: any individual unit of type, i.e., letters, figures, punctuation marks, or the space between words.

Character count: total number of characters in a line, paragraph, or an entire block of copy.

China marker: also called a grease pencil, a pencil with a colored waxy center, used for marking glazed or shiny surfaces such as glass or plastic. It is usually covered with a spiral-wrapped paper that can be peeled since the wax center cannot be sharpened.

Chipboard: low-density board made from mixed wastepaper, used when strength and quality are not desired. It is often used for backing fragile artwork.

Chrome: a full-color photographic positive on transparent film.

Coated paper: paper coated with pigment, which improves printability and opacity. The finishes range from dull to very glossy.

Cockle finish: bond paper with a puckered finish and usually containing a high rag content.

Collage: units of paper, cloth, wood, photographs, etc., having contrasting textures and patterns, pasted together to create one composition.

Color break: a boundary that separates two colors on a mechanical. To color-break is to indicate such boundaries for the printer, usually on a vellum or tracing paper overlay covering the mechanical board.

Color matching system: a system of specifying colors using numbered color samples available in swatch books such as the Pantone Matching System (PMS).

Color separation: separating full-color art into the four process colors by photographically shooting or electronically scanning it through color filters.

Color swatch: a small example of color attached to the camera-ready art indicating the color to be matched in printing ink.

Color transparency: See Chrome.

Compass: an instrument in the shape of two legs hinged at the top, used for describing arcs or circles with a pen or pencil.

Comprehensive: general term for a faithful representation (created solely by hand-rendering techniques), showing type and illustrations in the positions they will have in the final artwork. An advanced comprehensive is rendered solely or in part using photographic techniques.

Contact print: a print made by placing a negative or positive in contact with sensitized paper, film, or a printing plate, creating a same-sized image.

Continuous-tone: describes an image containing gradient tones from white through gray to black.

Copy: in printing preparation, any type, photographs, illustrations, or prepared mechanicals. In design and typesetting, any typewritten manuscript.

Copy fitting: determining the amount of space it will take to set a manuscript in a specific type style and size.

C-print: a nick-name for a color print made on Ektacolor paper, manufactured by Kodak. This method of printing is expensive but is of very high quality.

Crawling: the contraction of ink or other medium on a surface, caused by lack of adhesion.

Crop marks: short, fine lines drawn on camera-ready material to indicate to the printer where an image should be trimmed.

Cropping: eliminating the unwanted parts of a piece of artwork in any phase of production in order to emphasize a desired part.

C-stat: using the same process and paper (Ektacolor) as in making a C-print, only done faster and achieving a lesser quality than a C-print.

Cutout lettering: Letterforms printed on self-adhesive acetate sheets or film, that are cut out and burnished onto the working surface.

Cyan: one of the four process ink colors. Also called process blue.

Debossing: producing a recessed image on a printed or unprinted surface.

Deckle edge: untrimmed, ragged edge of text, cover, and other fancy papers.

Descender: the part of a lowercase letterform that extends below the baseline.

Designer's colors (gouache): opaque watercolor paints (available in tubes) that dry flat without streaking.

Diagonal-line scaling: a method of determining the correct proportion of enlarged or reduced artwork by means of drawing a diagonal line through two opposite corners of a rectangle (containing the artwork), so that any new sized rectangle whose horizontal and vertical lines intersect this diagonal line will be in exact proportion to the original.

Die cutting: the use of sharp, steel rules to cut special shapes in paper, cardboard, cloth or other material.

Direct positive (D.P.): the result of a photographic process that creates an enlarged or reduced reproduction of a piece of original art, which maintains the same black-and-white relationship as that of the original.

Display type: any bold, decorative, or large type used for headlines or titles (as opposed to text type).

Dots (dot screen): individual elements of the printed surface formed by a halftone screen.

Drop-out type: type reversed out to white on a solid, halftone or screened-tint background.

Dry-mounting: a process by which photographs and other paper materials are mounted on cardboard, using a tacking iron, specially processed dry-mounting tissue, and a dry-mounting press. Heat, pressure, and adhesive-coated tissue are the bonding agents.

Dry-transfer type: type carried on sheets of film that is transferrable to a surface by burnishing.

Dummy: mock-up showing positioning of type and illustrations in the final piece. Also a three-dimensional form or a set of blank pages made up in advance to show the size, shape, and style of a piece to be printed.

Embossing: producing a raised image on a printed or unprinted surface. When it is done on blank paper, or without ink, it is called blind embossing.

Emulsion: a photosensitive coating on film, paper, or glass that reacts to light.

Exposure: in photography, the time and intensity of light allowed to act upon a light-sensitive coating or emulsion.

Face (typeface): a particular size or style of one letter, as distinguished from another.

Finish: See Comprehensive.

Finished art: original or prepared type or artwork ready to be photographed for reproduction.

First print: the reverse, or negative of an original piece of art provided to a photostat house (not to be confused with the direct-positive process).

First proof: the first proof pulled after manuscript copy has first been typeset, to be submitted to proofreaders to check for errors before a reproduction quality print is made. Also called a galley proof.

Flat art: art that is completed and suitably prepared for reproduction.

Flat tone: area of dot formation producing a single-tone value (without gradation), with the dots being equal in size and equally spaced. Also called a tint or screen tint.

Flop: to "turn over" an image in shooting; to reproduce the original image facing the opposite direction.

Flush right or left: indication to the typesetter to line up type vertically at the right or left within a column.

Folder: a single piece of paper possessing one or more folds with each section showing a complete page.

Folio: page number.

Font: the complete grouping of characters belonging to one size of typeface including capital letters, lowercase letters, numerals, and punctuation marks.

Format: a term used for the size, style, and layout specifications for any printed piece. Also the physical dimensions of a piece.

Four-color process: a method of reproducing full color, camera-ready copy by separating it into three primary colors (red, yellow, blue) and black, and then using four printing plates to achieve the effect of all the colors of the original.

French fold: a single sheet of paper economically printed on one side and then folded once horizontally and once vertically to produce a four-page folder.

Frisket: a stencil or mask cut out of paper or film or painted on with liquid masking fluid to block out or protect parts of artwork or printing plates.

Galley: See First proof.

Gang up: to print a number of different jobs, or multiples of the same job, on one sheet. Also refers to the assemblage of multiple photostats or photolith films (which contain type or artwork) on a single surface in order to reduce the cost of production of some advanced comprehensive-rendering.

Gatefold: a sheet or page that folds inward yet opens like a gate when unfolded.

Gouache: See Designer's colors.

Grain: in paper, the predominant direction taken by a majority of the fibers in any sheet. Tearing and folding is cleaner with the grain rather than against the grain.

Greek type: a body of type having no logical readability; also called dummy type. It is used primarily in preparing comprehensives to illustrate the basic type style and size of the one to be used in the printed piece. It is also a method of hand-indicating type in which shapes of letters are imitated without creating actual words (to "Greek in").

Grid: a flexible planning aid made up of planned or designed horizontal and vertical cross rules used for the layout of a printed page or publication. A grid helps to create order by indicating where design elements such as text, rules, headings, and artwork should align.

Gutter: blank inner margin of a page that is next to the binding. Also the blank space between columns of type.

Hairline: the finest line or space that can be reproduced.

Halftone: a reproduction obtained by photographing a piece of original artwork through a screen, producing an effect in which tonal values are represented by a series of tiny evenly spaced dots of varying size and shape.

Heading: one or more words set, for emphasis, in a different size and/or style than the body of text.

Headline type: the most ostentatious line of type in a printed piece.

Holding lines: red or black lines drawn on a mechanical indicating the exact area where halftones and screen tints are to be placed.

Hot stamping: method of printing in which heated type or a stamping die is pressed against a thin leaf of gold or other metallic pigment and then on paper, plastic, leather, or board, transferring the image or lettering in gold or in a color to the surface.

India ink: permanent black ink made of lampblack and a glue binder. Some kinds of India ink are waterproof.

Initial: a large letter used to initiate a body of copy. It is usually placed at the beginning of a chapter.

Italic: style of lettering that slants to the right. It is primarily used to give emphasis to a word or words.

Justify: to create lines of type exactly equal in length (flush left and flush right) by adjusting the space between letters and words.

Key: to code copy to its layout by means of symbols, such as letters or numbers.

Keyline: drawn outline of an area showing position and size of a panel, color tint, halftone, or line drawing.

Kid finish: a slightly textured finish used on a high-grade or bristol paper.

Kneaded eraser: soft, pliable eraser used in drawing and mechanical preparation that does not leave behind "crumbs" as do other erasers.

Kraft paper: sturdy brown paper made from sulfate pulp; commonly used for wrapping.

Laid paper: paper with a watermark of finely spaced parallel lines.

Lamination: a clear plastic film bonded by heat to a printed sheet to enhance the appearance of or to protect the sheet.

Layout: hand-rendered design for a printed piece. Basically, there are three types of layouts (in order of degree of finish): thumbnail sketches, rough layouts, and finished layouts.

Letter spacing: the space between letters.

Ligature: two or more characters joined on one body to improve the letter fit of difficult letter combinations or for aesthetic purposes.

Light box: a frame with a translucent glass or plastic surface under which a light source is placed to provide illumination for tracing drawings, assembling film materials, or opaquing negatives.

Line art: any copy that is suitable for reproduction without the use of a halftone screen.

Line conversion: process of rephotographing continuous-tone photographs with line film, reducing all values in the picture to tones of either black or white.

Linen tester: a small magnifying glass in a folding frame, widely used for examining negatives, plates, and printed material.

Line spacing: also called leading, the space between lines of type.

Logo: the depiction of a company or product in a special design or symbol, which is used as a trademark or signature.

Logotype: a corporate name or product, printed in the style with which the company has chosen to be identified. Logotypes often contain specially drawn letterforms.

Loupe: compact plastic magnifying device used to examine type, artwork, or transparencies.

Lowercase: the small letters in type.

Lucy: projector used to enlarge or reduce opaque artwork, type, layouts, photographs, or transparencies.

Machine-coated: describes paper that is coated on one or both sides while on the paper-making machine.

Magenta: one of the four process ink colors, also referred to as process red.

Margin: area generally left blank around type and/or illustrative matter (i.e., top, bottom, and sides).

Mark up: to clearly mark all instructions for the typesetter on the copy, all type specifications: type size, typeface, line measures, leading, whether to set flush left or flush right, justified or centered, paragraph spacing, and indentation.

Mask: paper, film, or other frisketlike material used to block out or protect selected sections of an image. See also **Frisket.**

Matte finish: dull paper finish, without gloss or luster.

Mean line: also called the waist line or the x-line, the imaginary line that delineates the tops of lowercase letters without ascenders.

Measure: in typesetting, the pica width of a line or column of type.

Mechanical: camera-ready assemblage consisting of all design and type elements pasted down in printing position on one or more art boards.

Middle tones: in continuous-tone and halftone copy, the tonal range between highlights and shadows (the grays).

Moiré pattern: the optical conflict between the ruling of the halftone screen and the dots and lines contained in the original, which produces undesirable patterns, such as when reproductions are created from halftone proofs.

Montage: a composite image made by placing several different images together.

Mounting: attaching or adhering flimsy artwork, layouts, or comprehensives onto stiff mounting board for presentation or for protection.

Negative: a photographic image on film or paper in which the tonal values of the original are reversed so that the dark areas appear light and vice versa.

Newsprint: a paper made from ground wood pulp and small amounts of chemical pulp. Used in printing newspapers.

Opacity: in reproduction, the characteristic of paper that prevents the type or image printed on one side from showing through to the other. Also the covering power of liquid media.

Opaque: to paint out areas on a negative that are not wanted on the finished plate.

Overlay: transparent or translucent sheet of film or paper hinged over artwork with tape and used for protection, specifications, or color separation.

Overprinting: double printing; printing an image over an area that already has been printed.

Pagination: the numbering of pages in sequential order.

Paper grade: categorizing of paper based on size, weight, and grain. Paper grade is usually defined in terms of use (i.e., bond, offset, book, newsprint).

Pebble finish: a textured finish on paper made up of fine embossed designs.

Perfect binding: a relatively inexpensive method of binding pages together, generally in a book, and fixing them to the cover with a flexible adhesive.

Photocopy or photostat: a duplicate photograph made from an original print, transparency, or piece of flat artwork. A photocopy is also the result of any image duplicated on a copying machine.

Photomontage: two or more photographs combined into one image; also called a composite photograph.

Pica: a measurement equal to approximately 1/6 of an inch. There are twelve points in a pica. Also used to indicate measurements on a layout or mechanical or to designate typewriter type that has ten characters per inch (as opposed to elite, which has twelve characters per inch).

Plate finish: a finish that gives paper a smooth, hard surface.

Ply: one of several layers of paper bonded together to make a thicker sheet.

P.M.S.: abbreviation for the Pantone Matching System, which is a manufacturer's name for a widely used color matching system.

Point: smallest unit of measurement in the pica system. There are twelve points in a pica. Type is also measured in points.

Polyester film: plastic film base with high dimensional stability, used for various photographic films and drafting materials; generally replaces cellulose acetate.

Portfolio: a sampling of artwork and printed pieces shown by an artist to a prospective customer or client. Also refers to the container or cover for artwork or printed pieces.

Positive: photographic prints on film or paper with light and dark areas corresponding to the light and dark areas on the original piece of artwork.

Posterize: the conversion of a continuous-tone image to two or more flat tones, achieved photographically or by hand-rendering.

Presentation book: a multi-ring, loose-leaf binder holding several two-sided acetate pages commonly used to display photographs, layouts, or drawings.

Pressure-sensitive: describes an adhesive coating that is protected by a backing sheet until it is used, which allows material to stick without moistening.

Printing paper: paper or board specifically formulated for use in printing and for other production applications.

Proportional scale: a tool used in determining the size of artwork when enlarging or reducing for reproduction.

Rag paper: a high-quality paper made either partly or entirely from cotton fiber.

Ragged right (or left): in type, lines of varying length vertically aligned on one side and thus unaligned on the other.

Register marks: small crosses or other marks placed on camera-ready copy to allow the registering (aligning) of overlays, negatives, printing plates, and so on.

Registration: accurate positioning of two or more negatives, printing plates, impressions, type, artwork, and so on, so that they align exactly with each other.

Reversal film: a special contact film that does not reverse the tone of the original (i.e., a positive remains a positive and a negative remains a negative).

Reverse type: type that drops out of the background and assumes the color of the paper.

Right-angle fold: in binding, this refers to two or more folds that are at 90-degree angles to one another.

Right-reading image: an image that reads from left to right.

Roman: refers to letterforms that appear upright (as opposed to italic). Also refers to a typeface based on early Roman letterforms.

Rough: a sketch, usually on tracing paper, giving a general idea of the size and positioning of various design elements.

Rubylith: manufacturer's name for a red masking film.

Rule: any line used in conjunction with type whether typeset or hand-rendered.

Runaround: type set to fit around a photograph, illustration, or other element of design.

Saddle-wire stitching: to fasten a booklet with staples through the middle fold of the sheets.

Same size (S.S.): indication to a photostat machine operator that what is required is a 100 percent or same-sized reproduction of the original.

Sans serif: without serif (or the short cross stroke at the end of a letter stroke).

Scaling: determining the proportion of art when enlarging or reducing for reproduction.

Score: to impress or indent on a stiff paper where a fold will occur to make creasing easier and cleaner.

Screen: a photographically made glass or film-based screen used in printing to convert continuous-tone copy to halftone copy. Screens are designated by the number of ruled lines they contain, from fifty lines per inch to five hundred lines per inch. The greater amount of lines, the sharper the print.

Screened positive: opposite of a screened negative. The term is used to differentiate it from continuous-tone positives.

Screen printing: also called Silk screening, a process that employs a stencil adhered to a fine-mesh screen. Ink is then forced through the uncovered areas and onto the paper by using a wood-handled rubber blade called a squeegee.

Screen ruling: the number of lines per inch on a contact screen or glass halftone screen.

Screen tint: a flat, unmodulated tint or tone achieved by stripping a piece of halftone film onto the plate-maker's negative.

Second print: a positive, containing the same relationship of black to white as the original. A first print is the negative; the second print is the positive.

Self cover: a cover of a booklet that has the same paper and weight as the inside text pages.

Serif: the short cross stroke at the end of a letter stroke.

Set solid: to set lines of type without leading or additional spacing between lines.

Shading sheets and screens: dot, line, and halftone patterns printed on clear transparent plastic sheets with pressure-sensitive adhesive on the back. The use of these sheets enables one to shoot the original art as a line shot with the shaded areas retaining the effect of a halftone.

Shooting copy: any copy to be photographed for reproduction. Also, the act of photographing copy for reproduction.

Slip sheeting: inserting blank sheets of paper between printed sheets to prevent the ink from offsetting onto the back of a previously printed sheet.

Solid: in printing, refers to areas that are completely covered with ink or areas that print 100 percent of a color. In typesetting, it refers to setting lines of type without any leading.

Solvents: liquids that can dissolve certain materials or can suspend a pigment in ink. Also, a thinner or a cleaner (such as rubber-cement thinner).

Spec: a short term for specifying, meaning to determine and write type specifications on the copy for the typesetter.

Spine: the binding side of a book (also backbone) connecting the front and back covers.

Spot varnish: to varnish in selected areas only. This procedure is used for aesthetic purposes, as well as for increasing the durability of often-handled pieces. See also **Varnish.**

Spread: two facing pages. Also called a double-page spread.

Squeegee: a rubber blade mounted in a broad wooden handle used in silk-screen printing to force ink through the mesh of the screen and onto the paper smoothly.

Stencil: a means of applying patterns with ink or paint. The stencil itself is cut from thin paper, frisket film, metal, or cardboard and placed on the surface while ink or paint is brushed or sprayed into the areas left open. Also refers to the "open" area on a silk screen through which ink is forced.

Stet: a proofreader's mark indicating that copy marked for correction should stand as it was before the correction was made.

Stock: any material used to receive a printed image (i.e., paper, board, foil, etc.).

Surprinting: line copy superimposed over screened copy on the same printing plate.

Taboret: a piece of furniture equipped with doors and shelves, used to hold a designer's tools and materials.

Tacking iron: small hot iron used to secure artwork and backing so that they remain registered in the dry-mounting process.

Text paper: a general term used to describe an antique, wove or laid finish book paper. It often has deckled edges and is usually used for decorative applications.

Text type: refers to the main body type; often smaller in size than fourteen points. Also called body type.

Thinner: clear liquids (solvents, diluents, oils) added to inks or to rubber cement to reduce viscosity or tack.

Thumbnails: small, rough layout sketches.

Tint: a screen tint, or a photomechanical reduction of a solid color by screening. Also the act of adding white to a color.

Tissue overlay: thin, translucent paper placed over artwork (mechanicals) for protection or to indicate color breaks or corrections.

Tooth: a characteristic of paper; a slightly rough finish, enabling it to take ink readily.

Trademark: a unique graphic device that identifies a product or the company that manufactures it.

Transfer methods: means of transferring an image to another surface by coating or covering the original on the back with charcoal or graphite, then placing the original, faceup, onto a surface. The original image is traced over with a sharp tool, leaving the image in graphite on the surface.

Transfer type: type carried on sheets that can be transferred to a surface by cutting out self-adhesive letterforms or by burnishing down pressure-sensitive lettering.

Trim size: the final size to which a sheet will be cut after printing.

Type family: the complete range of sizes and variations of a typeface design (i.e., bold, roman, condensed, expanded, etc.).

Type font: complete assortment of characters for one size of typeface, including capital letters, lowercase letters, figures, and punctuation.

Type gauge: a special ruler marked in elite and pica typewriter increments used for determining the depth occupied by groups of various sizes of type and leading (used for copy fitting).

Typesetter: typographer, the person who sets type. Also describes any device that sets type.

Typesetting: the composition of type by any means.

Type specification catalog; type specimen sheet: typesetter's catalog containing various typefaces in many different sizes and weights.

Typo: a typographic error made by the typesetter.

Uncoated paper: basic paper produced on the paper-making machine without the application of a coating.

Unjustified type: lines of type set either flush left or flush right, aligning vertically on one side and not on the other.

Upper- and lowercase (u/lc): indicates that the copy is to be set in a mixture of uppercase (capital) and lowercase (small) letterforms.

Uppercase: the capital or larger letterforms of a typeface.

Varnish: Thin, protective coating applied to a printed piece. A varnish can be either matte or glossy. Also refers to part or all of the vehicle in the ink.

Vehicle: the liquid ingredient in ink serving as the carrier of the pigment, and binding the pigment to the substrate (stock).

Vellum: toothy, cream-colored and absorbent paper. Also, a heavy tracing paper.

Velox print: high-quality screened photographic print used in preparing mechanicals. Veloxes are line art and can be photographed along with line copy, saving stripping costs.

Vignette halftone: a halftone in which the edges fade imperceptibly into the white of the paper.

Watercolors: paints consisting of pigments ground in a solution of gum arabic, water, and a plasticizer such as glycerin. Transparent watercolors are called aquarelle and opaque watercolors are called gouache, the latter having white pigment added to them.

Waxed paper: paper coated with wax mechanically, either by wet waxing or dry waxing.

Waxer (wax-coating machine): machine that applies, with rollers, a pressure-sensitive coating of wax to the backs of prints, proofs, or photostats. Artwork is then burnished to the working surface where it is securely adhered yet can be easily lifted and shifted to change its position.

Weight: lightness or heaviness of a typeface. Type varies from light to regular to bold or extra bold.

Window: clear rectangular or square panel in a litho negative halftone. Negatives are positioned or stripped into this window with tape.

With the grain: describing the directional character of the fibers in paper. Paper folds easier with the grain as opposed to against the grain.

Word spacing: adding space between words to fill out a line of type to a given measure.

Wove paper: an uncoated paper with a uniformly smooth surface.

Wrong-reading image: any image that is reversed left to right so it can be read correctly only in a mirror.

x height: the height of lowercase letterforms without ascenders or descenders.

Zipatone: a manufacturer's name for a series of screen patterns imprinted on plastic sheets that can be used to achieve tone on various kinds of artwork.

Bibliography

Ballinger, Raymond. *Layout and Graphic Design.* New York: Van Nostrand Reinhold, 1980.

Borgman, Harry. *Art and Illustration Techniques.* New York: Watson-Guptill, 1979.

Cardamone, Tom. *Advertising Agency* and *Studio Skills.* New York: Watson-Guptill, 1970.

Craig, James. *Designing with Type.* New York, Watson-Guptill, 1971.

———. *Graphic Design Career Guide.* New York: Watson-Guptill, 1985.

———. *Production for the Graphic Designer.* New York: Watson-Guptill, 1974.

Gates, David. *Graphic Design Studio Procedures.* New York: Lloyd-Simone, 1982.

———. *Lettering for Reproduction.* New York: Watson-Guptill.

Gray, Bill. *Studio Tips.* New York: Van Nostrand Reinhold, 1976.

Lem, Phillip. *Graphic Master 2.* Los Angeles: Dean Lem, 1974.

Marquand, Ed. *How to Prepare and Present Roughs, Comps, and Mock-ups.* New York: Art Direction Book Company, 1985.

———. *How to Prepare Your Portfolio.* Second ed.; rev. third printing. New York: Art Direction Book Company, 1983.

Pocket Pal. New York: International Paper Company, published annually.

Sanders, Norman. *Graphic Designer's Production Handbook.* New York: Hastings House, 1982.

Stone, Bernard, and Arthur Eckstein. *Preparing Art for Printing.* New York: Van Nostrand Publishing Company, 1965.

Trademark Information

Agfa-Gevaert Repromaster 2001 is a trademark of AGFA-GEVAERT, INC.

Badger Propel is a trademark of BADGER AIR-BRUSH CO.

Berol Prismacolor Art Marker and Berol Verithin Colored Pencil are trademarks of BEROL USA.

Ad Art, Admaster, Art Vel, Graphics, and 100% Rag Layout Marker Pad are trademarks of BIENFANG PAPER CO. (div. of Hunt Manufacturing Co.).

Liquitex is a trademark of BINNEY & SMITH, INC.

Aquatec is a trademark of BOCOUR ARTIST COLORS, INC.

Haberule is a trademark of ARTHUR BROWN & BRO., INC.

Cello-Tak is a trademark of CELLO-TAK MFG., INC.

Colorcast Flint Paper is a trademark of CHAMPION INTERNATIONAL CORP.

AD Marker and Chartpak Adjustable Burnisher are trademarks of CHARTPAK (of Times Mirror Co.).

Chromatec and Chromatec Chromaslick are trademarks of CHROMATEC INC.

Design Art Marker is a trademark of DIXON TICONDEROGA CO.

Dacron, Lucite, and Mylar are trademarks of E.I. DU PONT DE NEMOURS & CO., INC.

Ektacolor Paper and Kodalith are trademarks of EASTMAN KODAK CO.

Artgum Eraser, Higgins Waterproof Ink, and Higgins Black Magic Ink are trademarks of FABER-CASTELL CORP.

Pink Pearl Eraser and RubKleen Eraser are trademarks of FABER, EBERHARD.

Color-Aid Paper is a trademark of GELLER ARTIST MATERIALS, INC.

General's Sketching Pencil is a trademark of GENERAL PENCIL CO.

Grumbacher No. GR197 is a trademark of M. GRUMBACHER INC.

Hunt 22B Nib, B-4 Speedball Pen, and Speedball C-type Nib are trademarks of HUNT MANUFACTURING CO.

X-Acto Ball Burnisher and X-Acto Knife are trademarks of HUNT/X-ACTO, INC.

Webril Wipes is a trademark of THE KENDALL CO.

Koh-I-Noor Pencil, Koh-I-Noor Rapidograph, Pelikan Graphos, Pelikan Ink Eraser, Pelikan Waterproof Drawing Ink, and Rotring Rapidograph 150 are trademarks of KOH-I-NOOR RAPIDOGRAPH, INC.

Krazy Glue is a trademark of KRAZY GLUE, INC.

Krylon Crystal Clear Spray and Krylon Workable Fixatif Spray are trademarks of KRYLON (div. of Borden, Inc.).

LetraJet, LetraJet Air Marker Propellant, Letraset Dry-Transfer Lettering, Letraset Matte, and Letraset Spoon Burnisher are trademarks of LETRASET USA, INC.

Com-Art is a trademark of MEDIA CO., INC.

Fome-Cor is a trademark of MONSANTO CO.

Canson Vidalon No. 90 and No. 110 are trademarks of MORILLA, INC.

Bainbridge #172 is a trademark of NIELSON & BAINBRIDGE.

Paasche Model V, Paasche Model VI, and Paasche No. 2 Pressure Tank are trademarks of PAASCHE AIRBRUSH CO.

Pantone® Coated Color Paper, Pantone® Coated Color Paper Selector, Pantone® Color Overlay Selector, Pantone® Color Specifier, Pantone® Graduated Tone Paper, Pantone® Marker, and Pantone® Matching System are trademarks of PANTONE, INC. "Pantone" is a check-standard trademark for color reproduction and color reproduction materials.

Gillot 170 Nib is a trademark of PENTALIC CORP.

Tru-Edge is a trademark of PILCER ENTERPRISES, INC.

Pilot Metallic Marker, Pilot Permanent Marker, and Pilot Razor Point Marker are trademarks of PILOT PEN CORP. OF AMERICA.

Flex-Opaque, Dr. Ph. Martin's Bleed-Proof White, Dr. Ph. Martin's Dyes, Dr. Ph. Martin's Radiant Concentrated Water Colors, and Dr. Ph. Martin's Synchromatic Transparent Water Colors are trademarks of SALIS INTERNATIONAL.

Schaedler Precision Rule is a trademark of SCHAEDLER QUINZEL.

STABILayout Marker and Stabilo-Pen Marker are trademarks of SCWAN-STABILO USA, INC.

MarsMatic-700 Technical Pen, Mars-Plastic Drawing Eraser, and Staedtler Lumo Color Waterproof Marker are trademarks of J.S. STAEDTLER, INC. (subs. of Staedtler Mars GmbH).

Colorflex is a trademark of STEIG PRODUCTS.

Strathmore Paper is a trademark of STRATHMORE PAPER CO. (div. of Hammermill Paper Co.).

Thayer & Chandler Model A is a trademark of THAYER & CHANDLER, INC.

Marvy Marker is a trademark of UCHIDA OF AMERICA CORP.

Glide-Liner and Rubylith are trademarks of ULANO CORP.

Winsor & Newton Gum Water and Winsor & Newton Series 7 are trademarks of WINSOR & NEWTON, INC. (div. of Reckitt & Coleman PLC).

Zipatone Burnisher and Zipatone Spray Adhesive are trademarks of ZIPATONE, INC.

Micropore Paper Surgical Tape, Post-It Note Pads, Scotch Magic Transparent Tape, Scotch Positionable Mounting Adhesive, 3M Burnisher, 3M Colorant Kit, 3M Color Key, 3M Image 'n Transfer, and 3M Squeegee are trademarks of 3M CO.

Index

Acetates, 47–48
Acetone, 48
Acrylics, 44–45
Adhesives, 50–55
Advanced comprehensives, 13–17
  for annual report assignment, 141–44
  audience for, 14
  choice to render comprehensives or, 14
  cost of, 14
  importance of execution, 14–15
  for magazine advertisement assignment, 149–51
  oversize, 16
  for package design assignment, 130–32
  producing, 91–96
  as samples in portfolio, 16
  techniques for, 13–14
  in three-dimensional form, 16
  uses of, 15
Airbrushes, 33–34
Altering type
  altering letter spacing, 79–80
  altering line spacing, 78–79
  curving type, 80–81
American Institute of Graphic Arts (A.I.G.A.), 5
Anamorphic lens, 24–25
Annual report assignment, 135–44
Art boards, 47
Art Directors Club, 5
Art knives, 31–32

Beam compass cutters, 30
Black tape, 50
Blind embossing, 101
Bond paper, 10, 46
Booklet dummies, 87–90
  binding for, types of, 87–88
  grids, 88
  procedure for making, 88–90
  spreads, 88
Bristol board, 47
Brushes and color applicators, 32–33
Brush ruling, 61
Bullets, 26
Butting, 74–75

Burnishers, 36

China-marking pencils, 39
Chipboard, 47
Chisel point, 59–60
Chromatec transfer system, 109–10
Circle templates, 27, 29
Circular proportional scale, 26–27, 74
Clamp-on dispensers, 36
Cleaning aids, 56
Clear acrylic sprays, 55
Color airbrushing, 107–8
Color-coated paper, 46
Colored inks, 43
Colored loops, 67
Colored pencils, 10, 39
Color film overlays, 48
Color Keys. *See* 3M Color Keys
Color lettering, 69–70
Color markers, 45
Color marker lettering, 70
Color media, 43–45
Color specifiers, 36–37
Compasses, 30
Comprehensives/Comprehensive layouts (or "comps"), 10–13. *See also* Advanced comprehensives
- acetate wrapping for, 99–100
- for annual report assignment, 141
- cellophane wrapping, 97–99
- enhancing and protecting, 96–100
- finished layouts differentiated from, 11
- for magazine advertisement assignment, 149
- media used for, 13
- for package design assignment, 129
- paper used in, 12
- props for, 15–16
- protecting and enhancing, 81–83
- purpose of, 11
- as samples in portfolio, 17
- techniques used in, 12
- as term, 2, 4
- varnishing, 96–97

Copying machines, 23
Curve, 27
Cutout lettering, 70–71
Cutting tools, 31

Debossing, 101–4
- definition, 101

Demonstrations, 123–51
- annual report assignment, 135–44
- magazine advertising assignment, 143–51
- package design assignment, 123–32

Desktop dispensers, 36
Diagonal line scaling, 73–74
Display type, 67–71. *See also* Altering type
Dividers, 30
Double-sided transparent tape, 50
Drafting tape, 50
Drafting vellum, 49
Drawing, with liquid media, 61
Drawing boards, 20
Drawing instruments, 29–30
Drawing leads and lead holders, 38–39
Drawing pencils, 38
Dry-mounting tissues, 53–54
- procedure, 54

Dry-transfer sheets, 49
Dulling sprays, 55
Dummies, 4, 83–90
- booklet dummies, making, 87–90
- boxes, 85–86
- cylinders, 86
- definition, 83
- folding, 84–85
- making folder dummies, 86–87
- making package dummies, 85
- scoring, 84

Dusting brushes, 34
Dyes, 43–44

Electric erasers, 40
Ellipse templates, 27–29
Embossing, 101–4
Enlarging thumbnail sketches, 72–73
Equipment, 19–25
- chairs and tools, 19–20
- condensing and expanding lens, 24–25
- copying machines, 23
- drawing boards, 20
- exposure units, 23
- hair dryers, 22
- lamps, 21
- light boxes, 21
- loupe, 21–22
- lucy, 22–23
- opaque projectors, 24
- photostat machines, 23–24
- rotating trays, 21
- taborets, 20–21

Erasers, 39–40
Exposure units, 23

Finished layouts, 9–10
- for annual report assignment, 139–40
- level of finish in, 9
- for magazine advertisement assignment, 147–48
- for package design assignment, 126

Finishes, 2, 4
Flat bastard file, 35–36
Flat sketching pencils, 39
Flexible curves, 29
Fome-Cor boards, 47
French curves, 29
Frisket paper, 46

Gouache, 43
Graphic designer, role of, 1
Graphic design problems, steps in solving, 3–17
Graphic design process
- advanced techniques in, 91–114
- basic steps in, 1–2, 3–17
- basic techniques in, 59–90

Graphite transfer method, 72
Grease pencils, 39
Grid, creating, 114
Gum erasers, 40

Hair dryers, 22
Household cement, 53

Illustration board, 12, 47
India inks, 42–43
Ink lettering, 69
Ink loops, 66–67
Inks, 42–43
Interim mechanicals, 77–78

Joining materials, 74–77

Kneaded erasers, 40

Layout and visualizing paper, 10, 46
Layouts. *See also* Finished layouts; Rough layouts
- dummies, 4, 83–90
- finished, 9–10
- "Greeked in," 8
- for presentations, 117–19
- rough, 7–9

Lead holders, 38–39
Leading, as term, 66
Lead pointers, 35
LetraJet, 34
Lettering and drawing pens, 40–41
Lettering brushes, 33
Lettering templates, 27
Letter spacing, altering, 79–80
Light boxes, 21
Light box tracing, 72
Lightweight cellophane, 48
Lightweight tracing paper, 46

Line copy, 13
Line spacing, altering, 78–79
Liquid media, drawing with, 61
Loupe, 21–22
Lucite rollers, 36
Lucy, 22–23
  advantages of, 139

Magazine advertising assignment, 145–51
Markers, 10, 45
Masking tape, 49–50
Mass-market oriented package design exploratory, 4
Mat boards, 47
Materials and media, 38–57
  acetates and film overlays, 47–49
  acrylics, 44–45
  adhesives, 50–55
  China-marking pencils, 39
  colored pencils, 39
  color media, 43–45
  designers' colors (gouache), 43
  drawing leads and lead holders, 38–39
  drawing pencils, 38
  dyes, 43–44
  erasers, 39–40
  inks, 42–43
  papers and pads, 45–47
  pencils, 38–39
  pens, 40–42
  presentation materials, 56
  spray coatings and fixatives, 55–56
  tapes, 49–50
  technical pens, 41–42, 43
  wipes and cleaning aids, 56
Matting
  definition and description, 81
  steps in, 82–83
Mechanical enlarging, 73
Metallic foil papers, 46
Metallic markers, 45
Mounting, definition, 81
Mounting boards, 47

Nonreproducing pencils, 39
Note pads, 37

Opaque projectors, 24
Opaquing Color Keys, 106–7
Opaquing whites, 45

Package design assignment, 123–32
Pantone Color Specifier, 37
Papers and pads, 45–46
Patterned tapes, 50
Pencil- and lead-sharpening tools, 35
Pencil erasers, 39
Pencil lettering, 67–69
Pencil loops, 66
Pencils, 38–39
Pencil sharpeners, 35
Pens, 40–42
Photocopies, for advanced comprehensives, 91–93
Photographic enlarging, 73
Photoliths, for advanced comprehensives, 94
Photostat machines, 23–24
Photostats, for advanced comprehensives, 93–94
Picture services and sources, 5
Plastic erasers, 40
Portable pencil sharpeners, 35
Portfolio presentation cases, 119–21
  advantages and disadvantages, 121
  multiring presentation books, 120
  multiring presentation cases, 120
  rigid presentation cases, 120–21
Post-It note pads, 37
Power burnishers, 36
Presentation materials, 56
Presenting designs, 115–21
  choosing pieces to present, 116
  choosing presentation vehicle, 119–21
  formats, determining, 119
  importance of, 115
  preparing for presentation, 116–19
Pre-Val Sprayer, 34–35
Printed papers, 46
Props, in comprehensives, 15–16
Protecting sprays, 55–56

Research and analysis, 4–6
  for annual report assignment, 135
  competition and environment of product, 4
  for magazine advertisement assignment, 145
  for package design assignment, 124
  at public library, 5
  questions to ask, 5
Rough, as term, 2, 3
Rough layouts, 7–9
  distinction between finished layouts and, 9
  for magazine advertisement assignment, 146
  for package design assignment, 125
  tools and media for, 9
Rubber cement, 50–52
  dispenser for, 52
Rubber-cement pickup, 52
Rubber-cement thinner, 52
Rub-down transfers, 108–10
Ruled colored lines, 64–65
Ruled ink lines, 64
Ruled pencil lines, 63–64
Ruling pens, 30–31

Sandpaper blocks, 35
Scaling artwork, 73–74
Schaedler Precision Rules, 26
Scissors, 32
Screen printing, 110–12
Seamless background paper, 46–47
Shading film overlays, 49
Silk screening, 110–12
  steps in, 111–12
Simulated printing techniques, 101–12
  embossing, 101–4
    definition, 101
  silk screens, 110–12
  3M Image 'N Transfer (I.N.T.) system, 104–6
Single-edge razor blades, 32
Sketches, for presentation, 117–19
Special effects, 112–14
Spray adhesive, 53
Spray coatings and fixatives, 55
Spray paints, 45
Squaring up, 60
Straightedges, 25
"Super comp." *See* Advanced comprehensives
Super glue, 53

Tape dispensers, 36
Tapes, 49–50
Technical pens, 41–42
  ink for, 43
Templates, 27
  making, 61
Text type, 61–67. *See also* Altering type
  ascenders, 66
  descenders, 66
  indication guidelines, 62
  loop method of indicating, 65–66
  rendering type on acetate overlays, 71
  ruled-line method of indicating, 63–65
  transferring techniques, 71–72
  type specimen sheet, 66
3M Color-Keys
  for advanced comprehensives, 94–96
  opaquing, 106–7
3M Image 'N Transfer (I.N.T.) system, 104–6
3M Magic Transparent Tape by Scotch, 50
3M Positionable Mounting Adhesive (PMA), 54–55

Thumbnail sketches, 6–7
- for annual report assignments, 137
- enlarging, 72–74
- for magazine advertisement assignment, 145
- need for several, 9
- for package design assignment, 124

Tools, 25–37
- airbrushes, 33–34
- art knives, 31–32
- beam compass cutters, 30
- brushes and color applicators, 32–33
- burnishers, 36
- circle templates, 27, 29
- circular proportional scales, 26–27
- clamp-on dispensers, 36
- color specifiers, 36–37
- compasses, 30
- cutting tools, 31
- desktop dispensers, 36
- dividers, 30
- drawing guides, 27
- drawing instruments, 29–30
- dusting brushes, 34
- ellipse templates, 27–29
- flat bastard file, 35–36
- flexible curves, 29
- French curves, 29
- lead pointers, 35
- LetraJet, 34
- lettering brushes, 33
- lettering templates, 27
- lucite rollers, 36
- manual burnishers, 36
- note pads, 37
- other color specifiers, 37
- Pantone Color Specifier, 37
- parallel ruling straightedges, 25
- pencil- and lead-sharpening tools, 35
- pencil sharpeners, 35
- rulers, 25
- rules and scales, 25
- ruling pens, 30–31
- sandpaper blocks, 35
- Schaedler Precision Rules, 26
- scissors, 32
- single-edge razor blades, 32
- tape dispensers, 36
- triangles, 26
- true-edge, 25–26
- T square, 26
- type gauge, 26
- utility knives, 32
- watercolor brushes, 33

Tracing paper
- advantage of, 10
- disadvantages of, 9
- lightweight, 46

Triangles, 26
True-edge, 25–26
T squares, 26
Type gauge, 26

Utility knives, 32

Vellum, 46
Visual estimation, 73
Visualizing paper, 10, 46

Water-based markers, 45
Watercolor brushes, 33
Wax, 53
White glue, 53
White paper tape, 50
Window mats, 10, 11
Wipes, 56